ZEITSCHRIFT FÜR ALLGEMEINE MIKRO-BIOLOGIE

AN INTERNATIONAL JOURNAL ON MORPHOLOGY, PHYSIOLOGY, GENETICS, AND ECOLOGY OF MICROORGANISMS

VOLUME 24 · 1984 · NUMBER 4

AKADEMIE-VERLAG · BERLIN

ISSN 0044-2208 Z. allg. Mikrobiol., Berlin 24 (1984), 4, 217–280 EVP 20,– M

Instructions to Authors

1. The journal publishes original papers, short communications, and review articles. Submission of a paper implies that it has not been published or has not been submitted for publication elsewhere. Manuscripts should be sent in duplicate with one set of the original illustrations to the Editor-in-Chief: Prof. Dr. U. Taubeneck, DDR-6900 Jena, Beutenbergstr. 11.

2. Manuscripts should preferably been written in English but may also be submitted in German. They should be typed double-spaced and be accompanied by a title page comprising: name and address(es) of the institution(s) where the work was done, title of the paper, and the complete name(s) of the author(s). Each paper must begin with a brief summary in English. Original papers should be divided into sections headed: Introduction, Materials and Methods, Results, Discussion, Acknowledgements, and References. A short title for use as running head should be provided. The exact mailing address of the author to whom correspondence, reprint requests *etc.* are to be addressed must be given at the end of the paper.

3. Tables, illustrations, and descriptive legends of the illustrations must be submitted on separate sheets. Each table should have a heading. The size of illustrations should not exceed the maximum printing area of 12 cm × 19 cm or 4.7 inches × 7.5 inches, respectively.

4. Literature citations in the text should be by author and year of publication. If there are more than two authors, only the first should be named, followed by "*et al.*". References should include only publications cited in the text. They should be given in alphabetical order:

   a) Books: Family name(s) and initials of author(s), year of publication. Title of the book. Volume and Edition. Publisher and place of publication, e. g.:

   BERTHELIN, J., BELGY, G. and MAGNE, R., 1977. Some aspects of the mechanism of solubilization and insolubilization of uranium from granites by heterotrophic microorganisms. In: Bacterial Leaching Conference (W. SCHWARTZ, Editor), pp. 261 to 270. Verlag Chemie Weinheim.

   b) Journals: Family name(s) and initials of authors(s), year of publication. Title of the paper. Abbreviated name of the journal, volume (underlined), number of the first and last page, e.g.:

   KÄPPELI, O., MÜLLER, M. and FIECHTER, A., 1978. Chemical and structural alterations at the cell surface of *Candida tropicalis*, induced by hydrocarbon substrate. J. Bacteriol., **133**, 952–958.

5. Galley proofs will be sent to the author in duplicate. One of them should be corrected and returned to the Editor-in-Chief or to Redaktion der Zeitschrift für Allgemeine Mikrobiologie, DDR-6900 Jena, Beutenbergstr. 11, as soon as possible.

6. 100 reprints of each paper are provided free of charge. Additional reprints may be purchased at cost price.

ZEITSCHRIFT FÜR ALLGEMEINE MIKRO-BIOLOGIE

AN INTERNATIONAL JOURNAL ON MORPHOLOGY, PHYSIOLOGY, GENETICS, AND ECOLOGY OF MICROORGANISMS

VOLUME 24 · 1984 · NUMBER 4

AKADEMIE-VERLAG · BERLIN

Die Zeitschrift für allgemeine Mikrobiologie soll dazu beitragen, Forschung und internationale Zusammenarbeit auf dem Gebiet der Mikrobiologie zu fördern. Es werden Manuskripte aus allen Gebieten der allgemeinen Mikrobiologie veröffentlicht. Arbeiten über Themen aus der medizinischen, landwirtschaftlichen, technischen Mikrobiologie und aus der Taxonomie der Mikroorganismen werden ebenfalls aufgenommen, wenn sie Fragen von allgemeinem Interesse behandeln.

Zur Veröffentlichung werden angenommen:

Orginalmanuskripte, die in anderen Zeitschriften noch nicht veröffentlicht worden sind und in gleicher Form auch nicht in anderen Zeitschriften erscheinen werden. Der Umfang soll höchstens 1½ Druckbogen (24 Druckseiten) betragen. Bei umfangreicheren Manuskripten müssen besondere Vereinbarungen mit der Schriftleitung und dem Verlag getroffen werden.

Kurze Orginalmitteilungen über wesentliche, neue Forschungsergebnisse. Umfang im allgemeinen höchstens 3 Druckseiten. Kurze Orginalmitteilungen werden beschleunigt veröffentlicht.

Kritische Sammelberichte und Buchbesprechungen nach Vereinbarung mit der Schriftleitung.

Terms of Subscription

Orders for the journal can be sent

- *in the GDR*: to a book-shop, or to the Akademie-Verlag, DDR-1086 Berlin, Leipziger Str. 3–4;
- *in the other socialist countries*: to a book-shop for foreign language literature or to the competent news-distributing agency;
- *in the FRG and Berlin (West)*: to a book-shop or to the wholesale distribution agency Kunst und Wissen, Erich Bieber, OHG, Wilhelmstr. 4–6, D-7000 Stuttgart 1;
- *in the other Western European countries*: to Kunst und Wissen, Erich Bieber GmbH, Dufourstr. 51, CH-8008 Zürich;
- *in other countries*: to the international book-and journal-selling trade, to Buchexport, Volkseigener Außenhandelsbetrieb der DDR, DDR-7010 Leipzig, Postfach 160, or to the Akademie-Verlag, DDR-1086 Berlin, Leipziger Str. 3–4.

Zeitschrift für allgemeine Mikrobiologie

Herausgeber: Im Auftrag des Verlages von einem internationalen Wissenschaftlerkollektiv herausgegeben.
Verlag: Akademie-Verlag, DDR-1086 Berlin, Leipziger Straße 3–4; Fernruf: 2236229 oder 2236221; Telex-Nr.: 114420; Bank: Staatsbank der DDR, Berlin, Kto.-Nr. 6836-26-20712.
Chefredaktion: Prof. Dr. Udo Taubeneck,
Anschrift der Redaktion: Zentralinstitut für Mikrobiologie und experimentelle Therapie der Akademie der Wissenschaften, DDR-6900 Jena, Beutenbergerstr. 11; Fernruf: Jena 852487; Telex-Nr.: 058621.
Veröffentlicht unter der Lizenznummer 1306 des Presseamtes beim Vorsitzenden des Ministerrates der Deutschen Demokratischen Republik.
Gesamtherstellung: VEB Druckerei „Thomas Müntzer", 5820 Bad Langensalza.
Erscheinungsweise: Die Zeitschrift für allgemeine Mikrobiologie erscheint jährlich in einem Band mit 10 Heften. Bezugspreis je Band 250,– M zuzüglich Versandspesen (Preis für die DDR 200,–M). Preis je Heft 25,– M (Preis für die DDR 20,– M).

Erscheinungstermin: Mai 1984.

Bestellnummer dieses Heftes: 1070/24/4

AN (EDV) 75218

Zeitschrift für allgemeine Mikrobiologie 24 (1984) 4, 219–222

(Institute of Chemistry "Boris Kidrič", Ljubljana, Jugoslavia,
and Institute of Chemistry of the Slov. Acad. Sci., Bratislava, CSSR)

Pichia labacensis n.sp.

ALEKSA CIMERMAN and ANNA KOCKOVÁ-KRATOCHVÍLOVÁ

(Eingegangen am 2. 8. 1983)

A new species of the yeast genus *Pichia*, isolated in a Yugoslav factory, is described. It differs from other *Pichia* species mainly in fermentation and assimilation abilities of various carbon substances. It is suitable for the production of fodder yeast from ultrafiltered spent sulfite liquor.

In studies of the possibilities of fodder yeast production from the permeate of ultrafiltered spent sulfite liquor, many different yeasts were tested. To find yeast strains able to utilize residual sugars in the substrate prepared from ultrafiltered spent sulfite liquor, we isolated cultures from samples taken at different places in a cellulose factory. Among all tested strains, the isolate designated K-100 was found most suitable. By using the methods of LODDER (1970), BARNETT and PANKHURST (1974) and KOCKOVÁ-KRATOCHVÍLOVÁ (1977), K-100 was determined to belong to the genus *Pichia*.

The yeast K-100 grows well in a substrate prepared from ultrafiltered spent liquor, yielding after 48 hrs of fermentation up to $8\,g \cdot l^{-1}$ biomass containing 37% raw protein and reducing the COD (Chemical Oxygen Demand) value of the filtrate to $15{,}000\,mg\,O_2 \cdot l^{-1}$ (FRIEDRICH *et al.* 1983).

Latin diagnosis

In extracto malti cellulae elongatae, (3–7) × (6–16) μm, singulae et in catenas.

Cultura in agaro multi post unum mensem (20 °C) cremea et mollis, albida, cum margine lobato et piloso. Pseudomycelium arboreum abundat.

Asci rotundi, ellipsoidei atque elongati cum appendice. Ascosporae rotundae et ellipsoideae, 1–4 in asco.

Glucosum, galactosum, saccharosum et raffinosum fermentantur.

Glucosum, galactosum, melibiosum, melezitosum, raffinosum, saccharosum, mannitolum, ethanolum, glycerolum, ethandiolum et lysinum assimilantur. Kalli nitras non assimilatur.

In temperature 5 °C–42 °C crescit, temperatura optima 28–37 °C.

Ad crescentiam vitaminae externae non sunt necessariae.

Typus cultura: K-100 in Collectione Yugoslaviae, Ljubljana et CCY 39-43-1 in Collectione Bohemo-Slovaca, Inst. Chem., Bratislava, CSSR, deposita.

Description of *Pichia labacensis sp. n.*

After 3 days of growth on beer wort at 28 °C, cells are budding, ellipsoidal to elongated, single or in chains, 3–7 × 6–16 μm (average: 5.46 ± 0.15 and 9.14 ± 0.23 μm). Culture in liquid medium formed very early a thin, white, wrinkled, creeping pellicle and a sediment.

Colony and smear on wort agar are white to cream coloured, dull. On onion agar tree-like pseudomycelium was formed.

The activity of sporulation was 24.5% on 0.5% acetate agar. Asci were ellipsoidal, spherical, elongated with linear or rhomboedric arrangement of spores. 1—4 spores in ascus were observed (12.5% 2-spored, 6.5% 3-spored, 5.5% 4-spored asci). Asci possessed an appendage. Ascospores spherical, ellipsoidal, with a brim.

Fermentation characteristics:

D-glucose +	Maltose —
D-galactose + (weak)	Lactose —
Sucrose +	Raffinose + (weak)

Assimilation characteristics:

D-glucose +	D-xylose —
L-fructose +	L-arabinose —
D-mannose +	Sucrose +
D-galactose +	Cellobiose —
L-sorbose —	D-glucitol —
Lactose —	D-galactitol —
Melibiose +	D-mannitol +
Melezitose +	D-ribose —
Raffinose +	*m*-inositol —
Trehalose —	Ethanol +
D-ribitol —	Glycerol +
i-erythritol —	Ethandiol + (weak)
Inulin —	Soluble starch —

Potassium nitrate is not assimilated, L-lysine is assimilated, but not DL-tryptophane. It grows on vitamine-free medium and in 60% sucrose. Urease is not produced, aesculin is weakly splitted. Discs containing 0.1 μg of actidione are not inhibitory but those containing 1 μg of actidione produce an inhibition zone of 8 mm diameter.

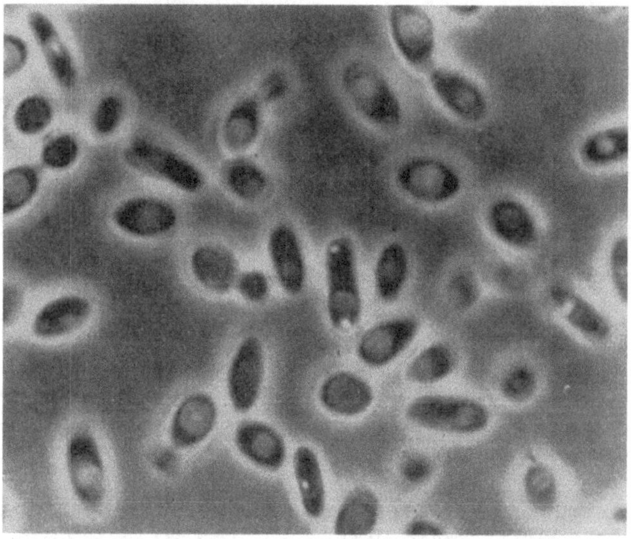

Fig. 1.
Cells in wort, 3 days, 30 °C, phase contrast, 500 ×

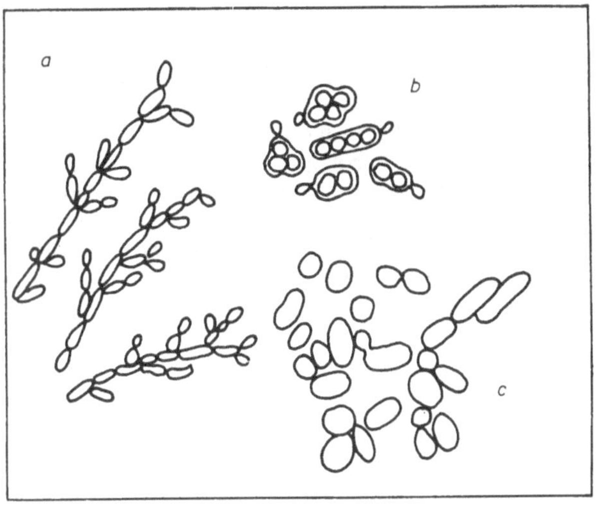

Fig. 2.
Growth and fructification: a — pseudomycelium type, b — asci with endospores, c — vegetative cells

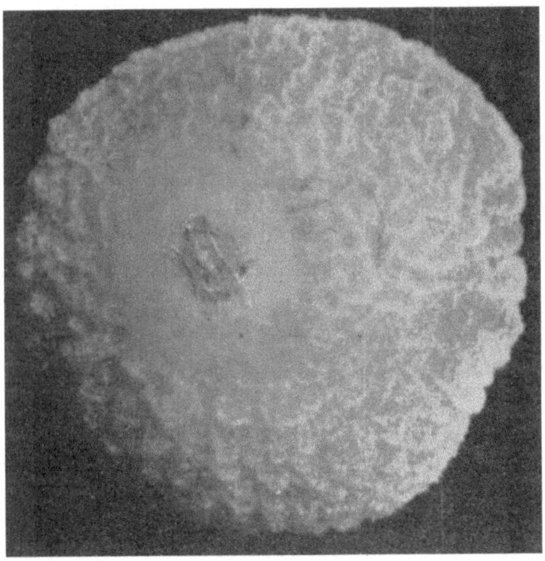

Fig. 3.
Giant colony on wort agar, 30 °C, 35 ×

It tolerates 8% ethanol in medium.

It grows at temperatures ranging from 5—42 °C, the optimal temperatures being 28—37 °C.

Type culture: Under sing. K-100 the strain is deposited in the Collection of Microorganisms, Ljubljana, Yugoslavia, and under CCY 39-43-1 in the Czechoslovak Collection of Yeasts, Inst. Chem., Bratislava, CSSR, (Figs. 1—3).

Discussion

According to the existing classification of all species described by LODDER (1970), including 36 species described after the year 1970, and using the system of KOCKOVÁ-KRATOCHVÍLOVÁ (1977), no species of the genus *Pichia* having the mentioned characteristics has been described. The fermentation principle (maltose—, sucrose +, lactose —) is common to 12 different *Pichia* species (*P. polymorpha*, *P. scolyti*, *P. guilliermondii*, *P. onychis*, *P. heinii*, *P. strasburgensis*, *P. nakazawae*, *P. etchellsii*, *P. amylophila*, *P. mississipiensis*, *P. veronae*, and *P. spartinae*). The nearest species is *P. spartinae* (AHEARN *et al.* 1970), but the new species differs from *P. spartinae* in some assimilation tests (Table 1). Other mentioned species differ from K-100 in more than three characters. Therefore, K-100 was described as a new species within the genus *Pichia* and named *Pichia labacensis*. The species name refers to the place of isolation of the strain.

Table 1

Comparison of assimilation tests in species *Pichia labacensis* and *Pichia spartinae*

Assimilation of	*P. spartinae*	*P. labacensis*
Maltose	+	—
Cellobiose	+	—
Trehalose	+	—
Raffinose	—	+

References

AHEARN, D. G., YARROW, D. and MEYERS, S., 1970. *Pichia spartinae sp. n.* from Louisiana marshland habitats. Antonie v. Leeuwenhoek, **36**, 503—508.

BARNETT, J. A. and PANKHURST, R. J., 1974. A New Key to the Yeasts. North Holland Publ. Comp. Amsterdam.

FRIEDRICH, J., CIMERMAN, A. and SUSA, L. 1983. Strokovno srečanje Biotechnologija, Meeting Abstracts, Ljubljana, p. 129.

KOCKOVÁ-KRATOCHVILOVÁ, A., 1977. Catalogue of Yeast Cultures. VEDA, Publ. House of the Slov. Acad. Sci., Bratislava.

LODDER, J. (Editor), 1970. Yeasts, a Taxonomic Study. North Holland Publ. Comp. Amsterdam.

Mailing address: RNDr. Dr. Sc. ANNA KOCKOVÁ-KRATOCHVILOVÁ, Institute of Chemistry of the Slov. Acad. Sci., 842 38 Bratislava, Dúbravská cesta, CSSR

Zeitschrift für allgemeine Mikrobiologie 24 (1984) 4, 223–230

(Akademie der Wissenschaften der DDR, Institut für technische Chemie, Leipzig, Direktor: Prof. Dr. sc. M. RINGPFEIL)

Growth of *Xanthomonas campestris* and xanthan accumulation

S. FIEDLER and U. BEHRENS

(Eingegangen am 15. 6. 1983)

During batch cultivations of *Xanthomonas campestris* different phases of growth have been observed. On glucose medium a short logarithmic phase is followed by a biochemical differentiation process. Its main result is nonbalanced growth and enhanced xanthan accumulation in the medium. The xanthan formation rate is influenced by the cell number at the end of the logarithmic growth. On the other hand, the amount of attainable xanthan is influenced by the growth rate during nonbalanced growth, which depends on the nitrogen concentration at the end of the logarithmic growth. When growing on glycerol, the cells display full logarithmic growth until the substrate is exhausted. The factor responsible for the biochemical differentiation and a detailed concept of xanthan accumulation are discussed. It is proposed that xanthan overproduction reflects typical secondary metabolism.

In previous papers (BEHRENS *et al.* 1977, 1980), a biphasic growth of the xanthan-accumulating bacterium *Xanthomonas campestris* has been reported. Phase I characterized by exponentially increasing cell numbers is followed by a phase of non-logarithmic growth (phase II) accompanied by extracellular xanthan accumulation. The kinetics of both growth and product formation correspond fairly well to those characteristic of the synthesis of secondary metabolites. A phase of reproductive growth without remarkable accumulation of metabolites in the medium (trophophase) is followed by a phase of limited growth and excretion of secondary metabolites (idiophase) (BU'LOCK 1964, DEMAIN 1971, CALAM 1979).

When *X. campestris* is cultivated in an appropriate medium, the two phases are clearly distinguished. A similar distinction, for instance, is characteristic of the growth of *Penicillium urticae* and patulin synthesis (BU'LOCK *et al.* 1965). However, the findings of STOUTHAMER (1977, 1979) exclude a generalization of this pattern, and he prefers to describe the transition as being more gradual.

The molecular basis which causes the biochemical differentiation is not yet fully understood. For *X. campestris* it is supposed that during exponential growth the increasing slime capsule limits the influx rate of nutrients from the medium into the cells, thereby leading to a decrease of the growth rate and an initiation of nonbalanced growth. This paper describes the relation of growth patterns to the synthesis of xanthan.

Materials and methods

Fermentation: The strain and methods of cultivation were described by BEHRENS *et al.* (1977), with the exception that peptone (3.5 g/l or 0.6 g/l nitrogen) served as the nitrogen source. Glycerol media contained 20 g/l glycerol instead of glucose as carbon source.

Analytical: Bacterial cells were counted. For gravimetrical biomass determination (dry matter), the samples were centrifuged. When necessary, the samples were diluted with water to decrease viscosity. The centrifuged cells were washed with 1/1000 N NaOH at room temperature to remove adhering xanthan slime. The determination of cell constituents was carried out with dried biomass preparations obtained by the same procedure. Protein was determined by the method of LOWRY *et al.* (1951). For RNA determinations the ribose moiety was measured photometrically after the removal of nucleotides (MEJBAUM 1939). DNA was determined in an analogous manner as deoxyribose (BURTON 1956). Commercial RNA resp. DNA from SERVA, Heidelberg, served as references. Total cellular carbohydrates were determined by the phenol-sulfuric acid method (DUBOIS *et al.* 1956) and extracellular xanthan by the quartolan method (BEHRENS *et al.* 1977). The soluble nitrogen content of cell-free broth was determined by the KJELDAHL-method. Glucose determinations were carried out enzymatically by using glucoseoxydase and peroxydase (commercial products of AW Dresden, see BEHRENS *et al.* 1977). Glycerol was determined colorimetrically after oxydation to formaldehyde and condensation with chromotropic acid (LAMBERT and NEISH 1957). The influence of residual glucose was found to be insignificant. Pyruvate was determined enzymatically by a modified colorimetric method (CZOK and LAMPRECHT 1974). To 1 ml of a 0.75 mol triethanolamine buffer (pH 7.6) containing 7.5 mmol EDTA, 0.1 ml 6 mmol NADH (AW Dresden), 0.10 ml of the sample, 1.9 ml bidistilled water and 0.02 ml L-lactatedehydrogenase (AW Dresden) were added.

Results and discussion

The cultivation of *X. campestris* in a medium containing glucose as carbon source and peptone as nitrogen source results in diauxic growth as measured by cell mass and cell

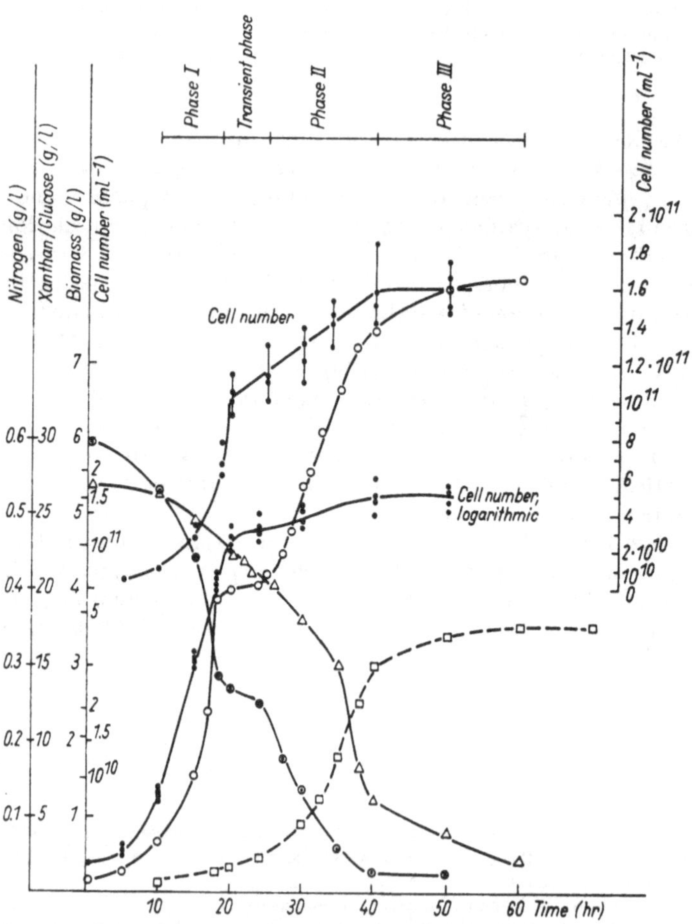

Fig. 1.
Growth and xanthan accumulation in a medium with glucose as carbon source and peptone as nitrogen source. ● cell number, ○ biomass, ⊙ nitrogen △ glucose, □ xanthan

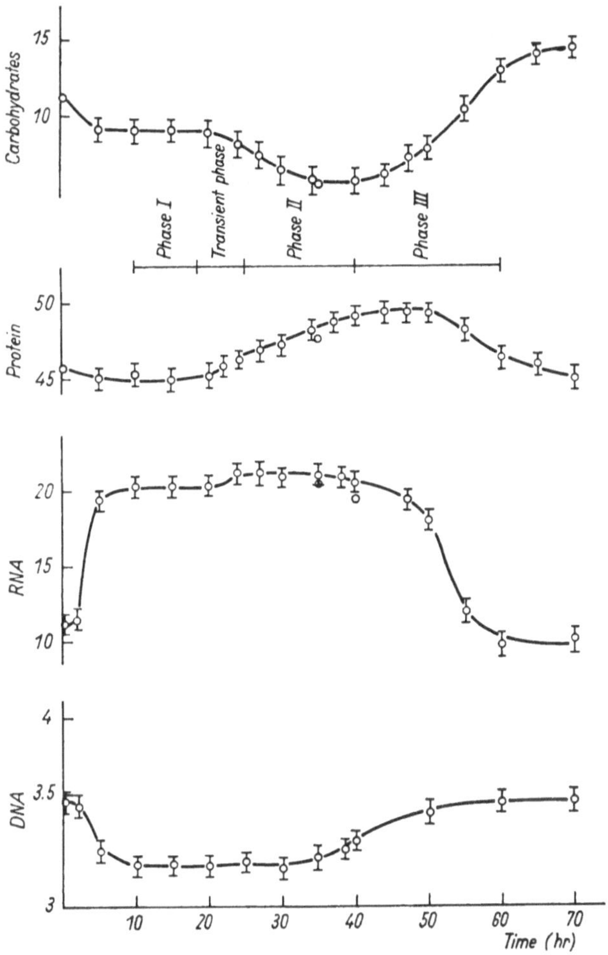

Fig. 2.
Variation of the macromolecular cell constituents during growth and xanthan formation. All data are expressed as percentages of dry cell mass

number (phase I and II in Fig. 1). Phase I displays mainly logarithmic growth (specific growth rate $\mu = 0.18\ hr^{-1}$) and is characterized by a high and constant RNA-content of the cells (Fig. 2).

Xanthan in small amounts is already secreted in phase I. By washing the cells with 1/1000 N NaOH, a polysaccharide, presumabely xanthan, precipitable with quartolan was recovered. With advancing fermentation time the polysaccharide slime layers around the cells increase, causing a reduction of diffusion rates of the nutrients into the microenvironment of the cells and by this a decline of the growth rate (Table 1). Growth imbalance is reflected by a reduction of protein/DNA and RNA/DNA quotients in phase II (Fig. 3). The transient phase displays a drastic reduction of the nitrogen uptake rate (Table 1). Contrary to this, the glucose uptake rate declines only insignificantly (Fig. 1), which corresponds to a nearly unchanged oxygen consumption. (Stottmeister unpublished). It may be assumed that cells in the transition phase undergo a process of biochemical differentiation for which energy must be made

Table 1
Specific growth rate μ and specific consumption rate of the nitrogen source $k_N = \frac{-dN}{dt} \cdot \frac{1}{X}$
Data were calculated from the curves of Fig. 1

time (hr)	10	12	14	16	18	20	22	24
μ	0.169	0.175	0.176	0.174	0.176	0.156	0.159	0.020
k_N	0.016	0.015	0.016	0.017	0.006	0.001	0.002	0.005

time (hr)	26	28	30	32	36	40	50
μ	0.05	0.045	0.041	0.038	0.033	0.029	0.004
k_N	0.005	0.003	0.003	0.002	0.002	0	0

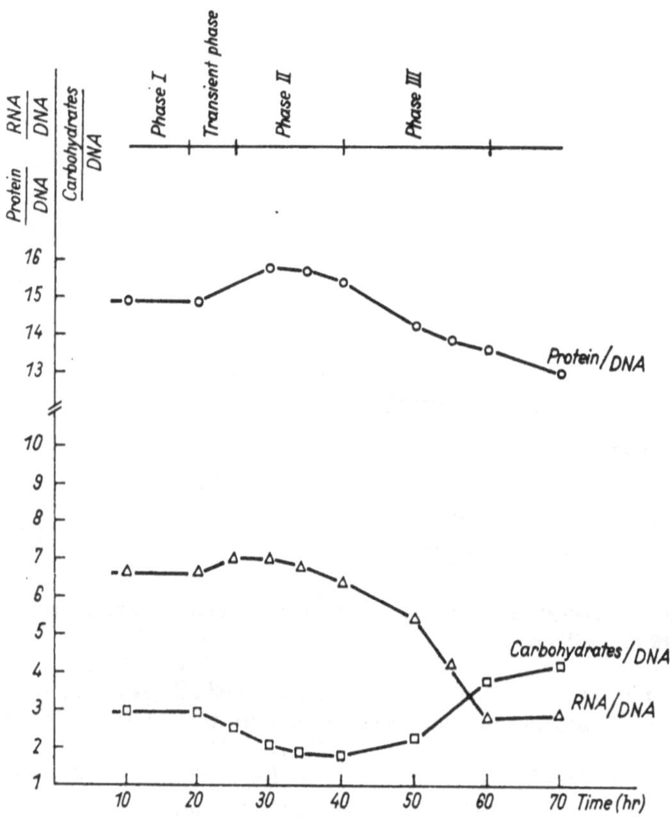

Fig. 3.
Variation of the macromolecular cell constituents during growth and xanthan accumulation. Data are related to the DNA content

available. The process of differentiation is further demonstrated by changing ratios of main constituents of the cells (Fig. 3). As could expected, in the transient phase the RNA content is not reduced, but rises slightly over the level attained during logarithmic growth. The rise in protein content follows with some temporal delay.

Similar cultivations performed with glycerol as the sole carbon source support the conception that a transitional stage is initiated by a xanthan layer developing around the cells. Glycerol as the sole carbon source cannot be used for xanthan synthesis, and the initiation of a transient phase was not to be expected. Accordingly in a glycerol medium undisturbed logarithmic growth proceeds until glycerol and nitrogen are almost completely exhausted. If glucose was added to a glycerol fermentation after exhaustion of the nitrogen source, no xanthan synthesis was initiated. Most of the glucose was respired to carbon dioxyde. In addition to this, small amounts of pyruvate and volatile fatty acids were synthesized (RÖMER 1979).

The end of the transient phase is characterized by an increasing rate of nitrogen uptake (Table 1). With some delay the rise of growth rate indicates the beginning of phase II (idiophase) which is characterized by nonbalanced growth (Fig. 2 and 3). The macromolecular components are subject to a perpetual change and the specific xanthan production rate in phase II is substantially higher than in phase I (Table 2).

Table 2

Specific xanthan production rates k_{xan} during phase II. $k_{xan} = \frac{d\text{xan}}{dt} \cdot \frac{1}{X}$

time (hr)	k_{xan}
25	0.17
30	0.18
35	0.16
40	0.13

The steady decline of the xanthan production rate in phase II does not impair the accumulation of xanthan which is counterbalanced by the increasing cell number. The exhaustion of the nitrogen source coincides with a drastic decline of the xanthan production rate and the end of replicatory growth, which is indicated by a rapid diminution of the RNA content. A further increase in cell mass is caused by reserve material formation, indicated by increasing amounts of carbohydrates (Fig. 2 and 3).

In this context the influence of the available nitrogen source, cell number and growth rates on xanthan production must be considered. As shown earlier (BEHRENS *et al.* 1980), the attainable amount of xanthan seems to be predetermined by the amount of the available nitrogen source at the beginning of nonbalanced growth. These findings have been further eludicated by fermentations started with glycerol as the sole carbon source. By feeding glucose when growth had reached specific stages defined

Table 3

Influence of available nitrogen on xanthan accumulation. Fermentation was described by BEHRENS *et al.* (1980) except that fermentation was started with glycerol (20 g/l). Glucose was added at specific stages of growth, indicated by the available nitrogen in the fermentation broth

available nitrogen (% of initial concentration)	maximal xanthan concentration (g/l)
0	0
11	0
26	2.7
41	5.3
55	8.8
58	9.0
76	13.6
89	17.2

by the amount of the available nitrogen in the fermentation broth, the amounts of xanthan obtained are related to the total amount of nitrogen and are independent of the cell number attained during logarithmic growth in glycerol broth (Table 3).

If a minimal amount of residual nitrogen was available, xanthan synthesis was initiated after the addition of glucose. Xanthan production coincides with the shift from logarithmic to nonbalanced growth. The results indicate that the quantity of xanthan produced depends on the extent of nonbalanced growth in phase II. The amount of cells produced during logarithmic growth in phase I seems to play a minor role.

If equal amounts of peptone and glucose were added to cultivations performed in media containing different concentrations of glycerol and peptone to obtain different cell numbers by logarithmic growth, the amount of xanthan produced was also found to be independent of the cell number attained at the end of the logarithmic growth (Table 4).

Table 4

Extent of logarithmic growth and xanthan accumulation. The bacteria were precultivated on a glycerol medium containing different quantities of peptone. When growth ceased, 20 g/l of glucose and 1.75 g/l of peptone were added

growth on glycerol		xanthan production
initial peptone concentration (g/l)	attained cell number (ml^{-1})	maximal xanthan concentration (g/l)
1.5	1.4×10^{11}	15.1
2.0	2.2×10^{11}	15.0
3.0	2.9×10^{11}	15.3

The results confirm the superior influence of the nonbalanced growth on the yield of xanthan. As observed in the experiments previously described, xanthan accumulation ceases when the nitrogen source is depleted.

In the course of phase II the cell mass is doubled, whereas the cell number, on the other hand, increases only 1.5 fold. A change in cell size has also been observed by MORAINE and ROGOVIN (1973).

The influence of nonbalanced growth on xanthan synthesis was further eludicated by a set of fermentations in which the overall concentrations of peptone remained constant, but the peptone concentration for growth in phase I was varied in order to obtain different cell numbers before the idiophasic growth and xanthan synthesis began (Table 5). By feeding peptone in such quantities as to provide the same overall

Table 5

Influence of overall nitrogen source (as peptone) and characterization of nonbalanced growth and xanthan formation of fermentations with different nitrogen concentrations. After exhaustion of the nitrogen source, peptone was added to provide the same overall concentrations

initial peptone concentration (g/l)	0.6	0.3	0.15
biomass at the end of phase I (g/l)	4.0	3.6	2.1
productivity of biomass (g/l h)	0.22	0.14	0.10
productivity of xanthan (g/l h)	1.2	0.6	0.35
xanthan yield (based on glucose)	0.65	0.61	0.58
max. xanthan concentration (g/l)	18.1	17.2	17.1
increase in cell numbers[1]) (fold)	1.6	1.54	1.63
increase in cell masses[1]) (fold)	2	2.1	2.2

[1]) based on data attained at the end of phase I

amount of peptone, growth and product accumulation in phase II differed substantially (Table 5). Surprisingly, the amounts of xanthan obtained at the end of each fermentation were nearly equal. Therefore, the influence of viscosity has to be taken into consideration. Peptone was fed during the postlogarithmic phase when the nitrogen source was nearly exhausted. Even in cultivations with media originally containing only small amounts of peptone, the additional feeding of peptone did not result in a resumption of logarithmic growth. Obviously, nonbalanced growth, once induced by a limited supply of a nutrient, cannot be reversed by shifting to nonlimiting substrate levels. Xanthan accumulation rates (productivity) are strongly influenced by the cell number attained in phase I, but the amount of xanthan in phase II up to the end of the fermentation is about the same (Table 5).

The attainable maximal xanthan concentration, therefore, seems to depend on the viscosity of the medium and on the fermenter system. It may strongly be influenced by agitation and/or the concentration of dissolved oxygen.

Considering these findings, it seems to be appropriate to discuss the concept of xanthan accumulation in greater detail. The formation of a slimy layer around the cells during the exponential growth has been demonstrated by washing the cells and cultivating on glycerol media, respectively. In glycerol media no xanthan can be formed and growth proceeds logarithmically until the nitrogen source is exhausted. Whether the xanthan layer causes an unspecific retardation of nutrient uptake or acts as a cationic exchanger, thus detaining specific nutrients, has not yet been established. Oxygen limitation may also play a dominant role.

Once logarithmic growth is stopped, an unknown trigger mechanism causes a biochemical differentiation. In this transient phase between trophophase (phase I) and idiophase (phase II) growth ceases and no nitrogen is consumed. However, glucose and oxygen consumption continue, indicating a need for energy. Nonbalanced growth characteristic of phase II is obviously a prerequisite for xanthan overproduction. On this basis xanthan overproduction seems to reflect typical secondary metabolism.

Increasing viscosity of the fermentation broth caused by extracellular xanthan accumulation reduces the diffusion rates. Thus the attainable xanthan concentration is also influenced by the fermenter system. Oxygen limitation may play a major role in the final stages of phase II.

References

BEHRENS, U., KLIMA, M. and FIEDLER, S., 1980. Growth and accumulation of polysaccharide by *Xanthomonas campestris*. Z. allg. Mikrobiol., **20**, 209–213.

BEHRENS, U., MAKARSKIJ, A., AMBROSIUS, J. und FRANKE, R., 1977. Stickstoffverwertung und Xanthanbildung bei *Xanthomonas campestris*. Z. allg. Mikrobiol., **17**, 339–345.

BU'LOCK, J. D., 1964. Aspects of secondary metabolism in fungi. Biogenesis Antibiot. Substances, Mater Panel Discussion, Congr. Antibiot. Prag (Publ. 1965), 61–71.

BU'LOCK, J. D., HAMILTON, D., HULME, M. A., POWELL, A. J., SMALLEY, H. M., SHEPHERD, D. and SMITH, G. N., 1965. Metabolic development and secondary biosynthesis in *Penicillium urticae*. Canad. J. Microbiol., **11**, 765–778.

BURTON, K., 1956. The conditions and mechanism of the diphenyl reaction for the colorimetric estimation. Biochem. J., **62**, 315–323.

CALAM, C. T., 1979. Secondary metabolism as an expression of microbial growth and development. Folia microbiol., **24**, 276–285.

CZOK, R. and LAMPRECHT, W., 1974. Methoden der enzymatischen Analyse (ed. by M. U. BERGMEYER). Verlag Chemie, Weinheim, Bergstr., **2**, 1491–1496.

DEMAIN, A. L., 1971. Overproduction of microbial metabolites and enzymes due to alteration of regulation. Adv. Biochem. Eng., **1**, 113–142.

DUBOIS, M., GILLES, K. A., HAMILTON, J. K., REBERS, P. A. and SMITH, F., 1956. Colorimetric method for determination of sugars and related substances. Anal. Chem., **28**, 350–356.

LAMBERT, M. and NEISH, C. A. 1950. Rapid method for estimation of glycerol in fermentation solutions. Canad. J. Research, **28** , 83–89.

LOWRY, O. H., ROSEBROUGH, N. J., FARR, A. L. and RANDALL, R. J., 1951. Protein measurement with the Folin method. J. biol. Chem., **193**, 265–275.

MEJBAUM, W., 1939. Über die Bestimmung kleiner Pentosemengen, insbesondere in Derivaten der Adenylsäure. Z. physiol. Chem., **38**, 117–120.

MORAINE, R. A. and ROGOVIN, P., 1973. Kinetics of the xanthan fermentation. Biotechnol. Bioeng., **15**, 225–237.

RÖMER, C., 1979. Untersuchungen zur Pyruvatausscheidung durch *Xanthomonas campestris* 2P. Diplomarbeit, Karl-Marx-Universität Leipzig, Sekt. Biowissenschaften.

STOUTHAMER, A. H., 1977. Penicilline na ongeveer 50 jaar onderzoek. Versl. Afd. Natuurk. Kon. Ned. Ac. v. Wet. **86**, 134–138.

STOUTHAMER, A. H., 1979. Energy Production, Growth and Product Formation by Microorganisms. Genet. Ind. Microorg., Proc. Int. Symp., 3rd 1978, (ed. by O. K. SEBEK and A. I. LASKIN). Am. Soc. Microbiol., Washington, D.C.

Mailing address: Dr. S. FIEDLER, AdW der DDR, Institut für technische Chemie, DDR 7050 Leipzig, Permoserstr. 15

Zeitschrift für allgemeine Mikrobiologie 24 (1984) 4. 231–237

(Institut für bakterielle Tierseuchenforschung der Akademie der Landwirtschaftswissenschaften der DDR, Jena, Direktor: Prof. Dr. sc. W. ERLER)

Beeinflussung von Virulenz und Immunogenität bei *Pasteurella multocida* durch Eisen *in vitro*

K.-D. FLOSSMANN, H. ROSNER, U. GRÜNKE[1]) und N. MIOSGA[2])

(Eingegangen am 11. 8. 1983)

The virulence of *Pasteurella multocida* strains in experimental infections of mice has been shown to depend on the iron contents of the cultivation media. Frequent passages of bacteria on iron-deficient growth media result in drastic decrease of virulence. In the case of one strain also immunogenicity was lowered. The results with nutritional media differing in their iron contents show that iron probably exhibits a regulatory effect on the production of a not yet identified virulence.

Virulenz und Immunogenität pathogener Bakterien spielen sowohl in der Auseinandersetzung Wirt-Parasit als auch bei der Entwicklung von Impfstoffen eine Rolle. Bei der Abwehr bakterieller Infektionen wird in den letzten Jahren dem Eisen große Bedeutung beigemessen (CHANDRA 1976, PAYNE u. FINKELSTEIN 1978, BULLEN *et al.* 1978). Ein Virulenzanstieg verschiedener bakterieller Erreger *in vivo* bei Eisenüberladung des Wirtes (JONES *et al.* 1977, STRAUSS 1978, FLOSSMANN u. MÜLLER 1980, FLOSSMANN *et al.* 1983a) bzw. eine Beeinflussung der Infektion durch den Eisenstatus des Wirtes (PUSCHMANN u. GANZONI 1977, HART *et al.* 1982) wurden experimentell beobachtet. In diesem Zusammenhang wird der Fähigkeit der Mikroorganismen zur Bildung sogenannter „Siderophore" Bedeutung beigemessen, um in der Auseinandersetzung Wirt-Parasit den essentiellen Mikronährstoff Eisen verfügbar zu machen (LUCKEY *et al.* 1972, NEILANDS 1981). Wenn es auch noch widersprüchliche Aussagen und nicht geklärte Phänomene hinsichtlich der Rolle der Siderophore gibt (ROGERS 1973, MILES u. KHIMJI 1975, KOCHAN *et al.* 1977, ODUGBEMI u. HAFIZ 1978), so wird doch die Fähigkeit zur Aufnahme des Eisens als ein wichtiger Faktor für die Virulenz der Mikroorganismen betrachtet (BULLEN *et al.* 1974, MAGNUSSON *et al.* 1979, FINKELSTEIN u. YANCEY, 1981, YANCEY u. FINKELSTEIN 1981, LEONG u. NEILANDS 1981).

Obwohl die Bedeutung von Eisen für Pasteurellen nachgewiesen wurde (HIGUCHI u. CARLIN 1957, FLOSSMANN *et al.* 1976, 1983a, FLOSSMANN u. MÜLLER 1980), ist über den Mechanismus der Aneignung des Eisens bei *Pasteurella multocida* bisher nichts bekannt. Ebenso ist nicht bekannt, wie sich definierte Änderungen des Eisenangebots *in vitro* auf die biologischen Eigenschaften der Bakterien auswirken. Im Rahmen von Untersuchungen zur Bedeutung des Eisens für die Abwehr von *P. multocida*-Infektionen interessierten wir uns für das Eisentransportsystem dieser Species. Um die Siderophorproduktion anzuregen, passagierten wir die Stämme *in vitro* in „low-iron"-Medien. Über die dabei gewonnenen Ergebnisse zur Bedeutung des Eisens für Virulenz und Immunogenität soll in dieser Arbeit berichtet werden.

---

[1]) VEB Keramische Werke Hermsdorf
[2]) Zentralinstitut für Mikrobiologie und experimentelle Therapie, Jena

Material und Methoden

Folgende *P. multocida*-Stämme wurden eingesetzt (in Klammern Serotyp nach NAMIOKA/CARTER sowie Wirtstier):
Institut für bakterielle Tierseuchenforschung Jena, DDR
35 (1/8:A; Kalb), 6417 (A, Schwein), 48 (A), 383 (6:B, Kalb), 383 *ts* (6:B; Mutante von 383);
National Institute of Animal Health, Tokio
VA 3 (5:A; Küken), Bunia II (6:E; Rind);
Central Public Health Laboratory, London
NCTC 10204 (9:D); 10324 (9:C; Reh), 10325 (9:D; Rind) 10326 (6:E; Rind);
Vet.-med. Forschungsinstitut „Pasteur", Bukarest, SRR
D 33 (2:D; Schwein), B 850 (2:B; Rind).

Die Herstellung der Nährmedien und Anzüchtung der Bakterien erfolgte nach HÖFER *et al.* (1982). Die Laborpassagen wurden in 5-ml-Kulturen durch Überimpfen von jeweils 10 Vol. % von 1—3 Tage alten Kulturen vorgenommen, Bebrütung bei 37 °C. Mäusepassagen wurden durch Infektion von NMRJ-Mäusen und nachfolgende Reisolierung der Keime aus den Organen (Milz, Leber, Lunge) vorgenommen.

Die Prüfung der Virulenz geschah im Mäuseversuch durch Bestimmung der Infektionsdosis, die 50% der Tiere tötet (LD_{50}).

Die Immunisierungsversuche mit $Al(OH)_3$-Totvakzine wurden in Anlehnung an ein Modell von CAMERON *et al.* (1978) an Mäusen durchgeführt. Bei Immunisierungen mit Lebendvakzine wurde dreimal im Abstand von 14 Tagen subkutan eine Dosis von 10^8—10^9 lebenden Bakterien verabreicht. 14 Tage nach der letzten Immunisierung wurde mit einer 24-h-Kultur (Dosis 1000—10000-fache LD_{50}) des Wildstammes 383 belastet und die Tiere 10 Tage beobachtet.

Die Eisenbestimmung erfolgte sowohl in den Nährmedien als auch in den Bakterien (nach Aufschluß mit 5%iger Trichloressigsäure; 1 h bei 100 °C) durch Atomabsorption mit dem Gerät AAS 1N (VEB CARL ZEISS, Jena).

Ergebnisse

Während Passagen in teildefinierten Nährmedien (Hefe-Extrakt-Medium 803 und 808, Tab. 4, 5) und chemisch definierten Medien mit „normalem" und erhöhtem Fe-Gehalt

Tabelle 1

Chemisch definiertes Nährmedium „902"[1])

Substanz	g/l H_2O (dest.)	mMol
$Na_2HPC_4 \cdot 12\ H_2O$	32,3	90
KH_2PO_4	1,36	10
NaCl	1,19	20
$MgSO_4 \cdot 7\ H_2O$	0,25	1
Glucose	6	30
L-Arginin · HCl	0,2	1
L-Asparaginsäure	1,6	12
L-Cystin	0,12	0,5
L-Glutaminsäure	0,15	1
DL-Serin	0,2	2
(oder L-Serin	0,1	1)
L-Isoleucin	0,065	0,5
L-Leucin	0,065	0,5
L-Phenylalanin	0,095	0,5
L-Tyrosin	0,09	0,5
Ca-Pantothenat	0,002	
Nicotinsäureamid	0,005	
Thiamin · HCl	0,001	
Orotsäure	0,015	0,1

[1]) „Fe-Defizit-Medium", dessen Fe-Gehalt (< 1 μg Fe/ml) nur durch die verwendeten Chemikalien bestimmt wird.

(Rezepturen 906 und 903, Tab. 4, 5) zu keiner Änderung der Virulenz der Bakterien führten, kommt es im Medium mit reduziertem Fe-Gehalt (Rezept 902, Tab. 1) zu einem starken Virulenzabfall der geprüften Stämme (Tab. 2, Abb. 1). Der Eisengehalt der Nährmedien liegt bei *ca.* 25 µg Fe/ml (903), 3–6 µg Fe/ml (803, 808), 2 µg Fe/ml

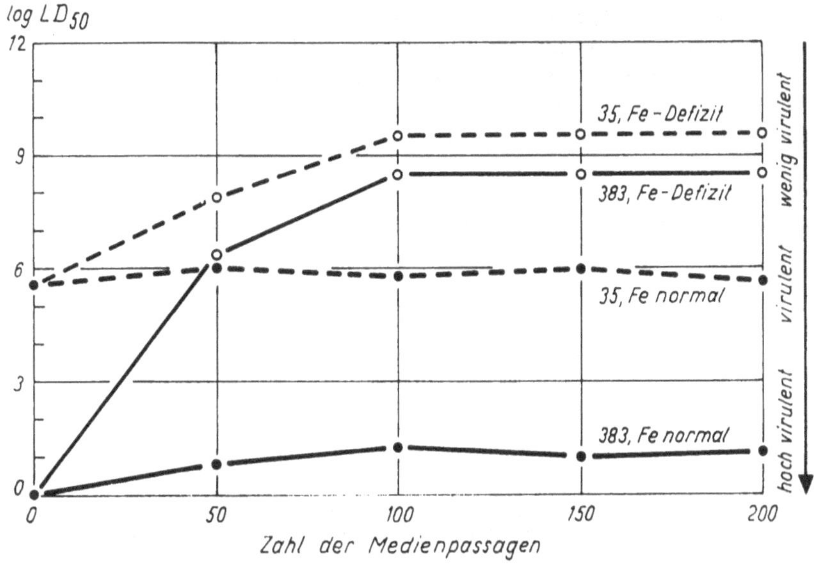

Abb. 1.
Einfluß des Eisengehaltes in einem chemisch definierten Nährmedium auf die Virulenz der *P. multocida*-Stämme 383 und 35 nach wiederholter Passagierung *in vitro*

Tabelle 2
Abhängigkeit der Virulenz (LD_{50}) vom Kulturmedium und von der Zahl der Passagen

Stamm	Zahl der Passagen —	50	100	150	200
a) Teildefiniertes Hefe-Extrakt-Medium 803					
383	1	2	28	17	5
35	6×10^5	1×10^6	3×10^6	5×10^5	$1{,}5 \times 10^4$
Bunia II	2	2	20		
D 33	32	19	25		
VA 3	95	1	40		
b) Chemisch definiertes Medium 906					
383	1	8	25	10	15
35	6×10^5	1×10^6	8×10^5	1×10^6	7×10^5
Bunia II	2	13	7		
D 33	32	45	28		
VA 3	95	1	16		
c) Eisenmangelmedium 902 nach Tabelle 1					
383	1	2×10^6	5×10^8	3×10^8	3×10^8
35	6×10^5	9×10^7	5×10^9	10^8	10^8
6417	25	$1{,}5 \times 10^3$	5×10^6	2×10^5	
B 850	13	9×10^5	3×10^8	10^8	
Bunia II	2	3	30	6×10^4	
D 33	32	10	4	5	10
VA 3	95	17	5	5	
10324	4×10^4	3×10^8	5×10^8	10^9	

(906) und < 1 µg Fe/ml (902). Der Eisengehalt des Mediums 902 (Tab. 1) wird nicht durch Supplementierung von Eisen, sondern durch die in ihm enthaltenen Substanzen (Spuren von Eisen!) hervorgerufen. 902 unterscheidet sich in seiner Zusammensetzung von 906 und 903 lediglich dadurch, daß dem Nährmedium kein zusätzliches Fe-III-citrat hinzugefügt wird. Die Medien 803 und 808 enthalten neben definierten Bestandteilen noch Hefe-Extrakt (Flossmann *et al.* 1977, Höfer *et al.* 1982).

Die Ergebnisse der Immunisierung mit dem Stamm 383 im Vergleich zu dem in seiner Virulenz durch Passagen in 902 abgeschwächten Abkömmling (383 Fe^{-}_{150} = 150 Passagen in Eisendefizitmedium 902) sind in Tabelle 3 zusammengefaßt. Die Immunogenität ist durch die Passagen stark beeinträchtigt.

Tabelle 3

Beeinflussung der Immunogenität von *P. multocida* durch Eisen-Defizit-Kultivierung *in vitro*

Vakzine	Zahl der Mäuse	Überlebende nach Infektions-belastung*)	% Überlebende
–	300	24	8
383 tot**)	300	178	59
383 lebend**)	120	117	98
383 Fe^{-}_{150} tot***)	140	20	14
383 Fe^{-}_{150} lebend***)	120	33	28

*) Infektionsbelastung mit 1000–10000 LD_{50} 14 Tage nach letzter Immunisierung
**) tot = $Al(OH_3)$-Adsorbatvakzine nach Cameron *et al.* (1978); lebend = Mutante 383 *ts*, Institut für bakterielle Tierseuchenforschung Jena
***) 150 Passagen des Stammes 383 im Fe-Defizit-Medium nach Tabelle 1

Die Virulenz der abgeschwächten Stämme (geprüft: 383 Fe^{-}_{200} und 35 Fe^{-}_{200}) kehrte durch Rückpassagen im Medium 906 bzw. in der Maus bis zur 25. Passage nicht zurück.

Tabelle 4

Fe-Gehalt von *P. multocida* (ng/mg Trockenmasse)

Stamm	Medium¹)			
	902	903	906	808
35	38	66	48	81
48	21	250	111	290
383	57	62	40	20
383 *ts*	22	47	85	19
6417	56	92	53	24
Bunia II	36	63	34	11
10204	21	32	22	10
10325	16	62	59	12
10326	13	12	9	11
∅	31 ± 16	76 ± 69	51 ± 31	53 ± 92

¹) Durchschnitt nach 20–50 Laborpassagen in den entsprechenden Nährmedien nach Flossmann *et al.* (1977), Höfer *et al.* (1982):
808 = fumarathaltiges teildefiniertes Hefe-Extrakt-Medium, Fe-Gehalt 3–6 µg/ml
906 = chemisch definiertes Medium mit Fe-III-citrat, Fe-Gehalt 1–2 µg/ml
902 = definiertes Medium nach Tabelle 1, Fe-Gehalt < 1 µg/ml
903 = definiertes Medium mit erhöhtem Fe-Gehalt (*ca.* 25 µg/ml)

Tabelle 5

Fe-Gehalt von *P. multocida* nach Passagierung in verschiedenen Medien

Stamm	Medium[1])	Fe-Gehalt in μg/mg Trockenmasse	
		nach 5–10 Passagen	nach *ca.* 150 Passagen
383	803	0,7	0,06
	906	0,1	0,02
	902	0,12	0,03
35	902	0,1	0,01
VA 3	803	0,06	0,05
	906	0,08	0,02
	902	0,1	0,03
D 33	803	0,5	0,17
	906	0,05	0,02
	902	0,15	0,02
Bunia II	803	0,09	0,1
	906	0,05	0,04
	902	0,15	0,09

[1]) Bezeichnungen wie in Tabelle 4,
803 = teildefiniertes Hefe-Extrakt-Medium, Fe-Gehalt 3–6 μg/ml

Der Eisengehalt der Bakterien differiert je nach Kultivierungsbedingungen (Tab. 4; und 5). In Medien höherer Fe-Konzentration (803, 808, 903) ist auch der Fe-Gehalt der Bakterien am höchsten (bis 0,2% der Trockenmasse!). Der Fe-Gehalt der Zellen schwankt insgesamt recht stark (0,001 bis 0,24%, durchschnittlich 0,011% der Trockenmasse) und sinkt nach hoher Passagenzahl meist ab. Der Fe-Gehalt der Bakterien, die im „Fe-Defizit-Medium“ 902 kultiviert wurden, unterscheidet sich nicht signifikant von in „Normalmedium“ 906 kultivierten Bakterien.

Diskussion

Obwohl wir bei *P. multocida*-Stämmen Eisentransportaktivitäten, die auf ein Siderophor hindeuten, nachweisen konnten (Veröffentlichung in Vorbereitung), ist es uns bisher nicht gelungen, ein Siderophor sicher zu charakterisieren. Die vorliegenden Ergebnisse von *in vitro*-Versuchen weisen jedoch nach, daß dem Eisen, wie schon bei *in vivo*-Experimenten (FLOSSMANN u. MÜLLER 1980, FLOSSMANN *et al.* 1983) festgestellt, für die Virulenz große Bedeutung zukommt. Auf einem Nährmedium mit einem verminderten Eisengehalt ist nach einer großen Anzahl von Passagen die Virulenz deutlich, offenbar irreversibel, erniedrigt. Dieser virulenzschwächende Effekt tritt bei den Stämmen nach unterschiedlicher Passagenzahl ein, was wiederum auf Stammesunterschiede hinsichtlich des Einflusses von Eisen zurückgeführt werden kann (Tab. 2). Der Fe-Gehalt der Bakterien liegt in dem für andere Bakterien bereits mitgeteilten Bereich um 0,01% der Trockenmasse (NEILANDS 1974). Obwohl er stark schwankt, kann der Eisengehalt in den Bakterien selbst keine entscheidende Ursache für die Virulenzänderung sein, da wir diesbezüglich keine Unterschiede nach Kultivierung in den beiden Medien 902 und 906 feststellen konnten (Tab. 4 und 5). Bakterien, die in dem „Eisendefizit-Medium“ 902 unter Virulenzabschwächung (Abb. 1) gewachsen waren, wiesen etwa den gleichen Fe-Gehalt auf wie jene, die in dem Medium 906 kultiviert wurden. Ein echtes „Eisen-Defizit“, bezogen auf den Fe-Gehalt der Bakterien, liegt damit nicht vor. Auf stark Fe-haltigen Medien gezüchtete Bakterien hatten einen hohen Eisengehalt (Tab. 5). Eine Korrelation von Fe-Gehalt und

Virulenz, wie sie ODUGBEMI und DEAN (1978) bei Neisserien feststellten, konnten wir bei *Pasteurella multocida* nicht nachweisen. Offenbar reguliert jedoch der Fe-Gehalt einen bisher unbekannten Virulenzfaktor, wie es z.B. bei Corynebakterien beobachtet wurde (EDWARDS u. SEAMER 1960). Ein solcher Faktor könnte z.B. ein Siderophor sein, dessen Produktion durch Fe stark beeinflußt wird (NEILANDS 1981).

Ein Widerspruch besteht offenbar zwischen unseren Ergebnissen und denen von WARING *et al.* (1953), nach denen bei Brucellen bei „Fe-Defizit" ein Virulenzanstieg festgestellt wurde. Die Ursache für diesen Widerspruch kann in der Natur der untersuchten Species liegen. Allerdings sind auch qualitative Aspekte hinsichtlich der Definition des Begriffes „Fe-Defizit" zu beachten. Sicher ist, daß Eisenbedarf und Eisenstoffwechsel trotz großer Analogien bei unterschiedlichen Mikroorganismen sehr variieren können. Ähnlich widersprüchliche Resultate ergeben sich bei der Wirkung des Eisens auf den Infektionsverlauf im Makroorganismus.

Die vorliegenden Ergebnisse haben insofern eine direkte praktische Bedeutung, da man mittels der Passagierung in definierten Medien bei entsprechender Wahl des Fe-Gehaltes Virulenz- und Immunogenitätsänderungen hervorrufen kann. Dies ist für Fragestellungen der Impfstofforschung von Bedeutung (FLOSSMANN *et al.* 1983). Die weitere Klärung des Phänomens und seiner Ursachen, besonders im Hinblick auf die Ausbildung von Siderophoren, ist Gegenstand einer nächsten Veröffentlichung.

Für experimentelle Hilfe und Unterstützung danken wir Frau CHRISTINE GRAJETZKI und Frau GISELA THIELE.

Literatur

BULLEN, J. J., ROGERS, H. J. and GRIFFITHS, E., 1974. Bacterial iron metabolism in infection and immunity. In: NEILANDS, J. B. (Ed.): Microbial Iron Metabolism. Academic Press London-New York, 517–551.

BULLEN, J. J., ROGERS, H. J. and GRIFFITHS, E., 1978. Role of iron in bacterial infection. Curr. Top. Microbiol. Immunol., **80**, 1–35.

CAMERON, C. M., ENGELBRECHT, M. M. and VERMEULEN, A. S. M., 1978. The serological and immunological relationship of type strains A and D of *Pasteurella multocida* to field isolates from sheep. Onderstepoort J. Vet. Res., **45**, 215–220.

CARTER, G. R., 1955. Studies on *Pasteurella multocida*. I. A haemagglutination test for the identification of serological types. J. Vet. Res., **16**, 481–484.

CHANDRA, R. K., 1976. Iron and immunocompetence. Nutr. Rev., **34**, 129–132.

EDWARDS, D. C. and SEAMER, P. A., 1960. The uptake of iron by *Corynebacterium diphtheriae* growing in submersed culture. J. Gen. Microbiol., **22**, 705–712.

FINKELSTEIN, R. A. and YANCEY, R. J., 1981. Effect of siderophores on virulence of *Neisseria gonorrhoeae*. Infect. Immun., **32**, 609–613.

FLOSSMANN, K.-D., FINSTERBUSCH, L., ERLER, W. und FEIST, H., 1976. Bedeutung von Eisen für die Ernährung von Pasteurellen und Salmonellen. Z. allg. Mikrobiol., **16**, 217–219.

FLOSSMANN, K.-D., ROHRMANN, B., FEIST, H., HÖFER, M. und ERLER, W., 1977. Chemisch definierte und teildefinierte Nährmedien für *Pasteurella multocida* und *Pasteurella haemolytica*. Arch. exper. Vet. med., Lepzig, **31**, 61–69.

FLOSSMANN, K.-D. und MÜLLER, G., 1980. Einfluß von Eisen auf die Virulenz von *Pasteurella multocida*. Acta biol. med. german., **39**, 327–334.

FLOSSMANN, K.-D., MÜLLER, G. und HEILMANN, P., 1983a. Wirkung parenteraler Eisengaben auf die experimentelle Infektion mit *Pasteurella multocida* bei Maus und Schwein. Arch. exper. Vet. med., Leipzig, **37**, 217–225.

FLOSSMANN, K.-D., ROSNER, H., FEIST, H., ERLER, W., HÖFER, M. und GRAJETZKI, CH., 1983b. Verfahren zur Kultivierung von Bakterien der Species *Pasteurella multocida*. DDR-WP., C 12 N/248 575 1, 8. 3. 83.

FLOSSMANN, K.-D., ROSNER, H., FEIST, H., ERLER, W. und HÖFER, M., 1983c. Verfahren zur Virulenzabschwächung bei Bakterien der Gattung *Pasteurella*, Species *P. multocida*. DDR-WP., C12 N/248 576 8, 8. 3. 83.

HART, R. C., KADIS, S. and CHAPMAN jr., W. L., 1982. Nutritional iron status and susceptibility to *Proteus mirabilis* pyelonephritis in the rat. Canad. J. Microbiol., **28**, 713–717.

HIGUCHI, K. and CARLIN, C. E., 1957. Studies on the nutrition and physiology of *Pasteurella pestis*. I. A casein hydrolyzate medium for the growth of *Pasteurella pestis*. J. Bacteriol., **73**, 122–129.

HÖFER, M., FLOSSMANN, K.-D., ERLER, W. und FEIST, H., 1982. Nährmedien zur Kultivierung von *Pasteurella multocida*. Arch. exper. Vet. med., Leipzig **36**, 565–575.

JONES, R. L., PETERSON, C. M., GRADY, R. W., KUMBARACI, T. and CERAMI, A., 1977. Effects of iron chelators and iron overload on *Salmonella* infection. Nature, **267**, 63–65.

KOCHAN, I., KVACH, J. T. and WILES, T. I., 1977. Virulence-associated acquisition of iron in mammalian serum by *Escherichia coli*. J. Inf. Dis., **135**, 623–632.

LEONG, S. A. and NEILANDS, J. B., 1981. Relationship of siderophore-mediated iron assimilation to virulence in crown gall disease. J. Bacteriol., **147**, 482–491.

LUCKEY, M., POLLACK, J. R., WAYNE, R., AMES, B. N. and NEILANDS, J. B., 1972. Iron uptake in *Salmonella typhimurium*: Utilization of exogenous siderochromes as iron carriers J. Bacteriol., **111**, 731–738.

MAGNUSSON, K.-E., KIHLSTRÖM, E., NORQVIST, A., DAVIES, J. and NORMARK, S., 1979. Effect of iron on surface charge and hydrophobicity of *Neisseria gonorrhoeae*. Infect. Immun., **26**, 402 to 407.

MILES, A. A. and KHIMJI, P. L., 1975. Enterobacterial chelators of iron: their occurrence, detection, and relation to pathogenicity. J. Med. Microbiol., **8**, 477–490.

NAMIOKA, S., 1970. Antigenic analysis of *Pasteurella multocida*. Nat. Inst. Anim. Health, **20**, 97–108.

NEILANDS, J. B., 1974. Iron and its role in microbial physiology. In: NEILANDS, J. B. (Ed.). Microbial Iron Metabolism. Academic Press London—New York, 3–33.

NEILANDS, J. B., 1981. Microbial iron compounds. Ann. Rev. Biochem., **50**, 715–731.

ODUGBEMI, T. O. and DEAN, B., 1978. Iron contents of different colonial types of *Neisseria gonorrhoeae*. J. Gen. Microbiol., **104**, 161–163.

ODUGBEMI, T. O. and HAFIZ, S., 1978. The effects of iron chelators on the colonial morphology of *Neisseria gonorrhoeae*. J. Gen. Microbiol., **104**, 165–167.

PAYNE, S. M. and FINKELSTEIN, R. A., 1978. The critical role of iron in host-bacterial interactions. J. Clin. Invest., **6**, 1528–1440.

PUSCHMANN, M. and GANZONI, A. M., 1977. Increase resistance of iron-deficient mice to *Salmonella* infection. Infect. Immun., **17**, 663–664.

STRAUSS, R. G., 1978. Iron deficiency, infections, and immune function: a reassessment. Amer. J. Clin. Nutr., **31**, 660–666.

WARING, W. A., ELBERG, S. S., SCHNEIDER, P. and GREEN, W., 1953. The role of iron in biology of *Brucella suis*. J. Bacterive., **66**, 82–91.

YANCEY, R. J. and FINKELSTEIN, R. A., 1981. Assimilation of iron by pathogenic *Neisseria spp.* Infect. Immun., **32**, 592–599.

Anschrift: Dr. sc. K.-D. FLOSSMANN, Institut für bakterielle Tierseuchenforschung Jena der AdL der DDR, DDR-6909 Jena 9, Naumburger Str. 96a

Zeitschrift für allgemeine Mikrobiologie 24 (1984) 4, 238

Buchbesprechung

T. D. BROCK, Membrane Filtration. A User's Guide and Reference Manual. XII + 381 S., 115 Abb. 26 Tab. Berlin—Heidelberg—New York 1983. Springer-Verlag. DM 87,00. ISBN 3-540-12128-5.

Mit dieser Publikation ist erstmals eine vollständige Darstellung der theoretischen Grundlagen der Herstellungsprinzipien und der Anwendungsmöglichkeiten der verschiedenen Typen von Membranfiltern verfügbar. Die Überschriften der 14 Kapitel lassen den umfassenden Charakter des Buches deutlich werden: Einleitung; Struktur und Funktion der Membranen; Herstellung von Membranfiltern; Charakterisierung und Standardisierung von Membranfiltern; Produzenten von Membranfiltern; Auswahl und Gebrauch eines Filtersystems; Filtration zur Sterilisation und in Produktionsprozessen; Anwendungen in Verbindung mit direkter Mikroskopie; Zählung lebensfähiger Keime; Mikrobiologische Untersuchung von Wasser; Biochemische und analytische Anwendungen; Virusbestimmung im Wasser; Ultrafiltration und umgekehrte Osmose; Luftfiltration. Es folgen ein Anhang mit den Filterprogrammen einiger Hersteller (GELMAN Sciences, MILLIPORE Corporation, NUCLEOPORE Corporation, SARTORIUS Filters Inc.) und eine weit über 200 Publikationen umfassende Liste von Referenzen, deren Schwergewicht auf dem angelsächsischen Schrifttum liegt. Der ausgezeichnete abschließende Sach-Index ermöglicht die schnelle Orientierung über spezielle Probleme.

Das Buch ist als Lehrbuch, als Grundlage für praktische Kurse und als Nachschlagewerk gleichermaßen geeignet. Der Text wird durch zahlreiche hervorragende Abbildungen eindrucksvoll ergänzt.

Im Vorwort erwähnt der Autor, daß dieses typographisch ausgezeichnet anmutende Buch mit Hilfe modernster Möglichkeiten („modern word processing equipment" und „computerized typesetting") produziert wurde, um die Herstellungszeit zu verkürzen. Leider ist nicht ersichtlich, welcher Zeitraum für die Herstellung tatsächlich benötigt wurde.

U. TAUBENECK (Jena)

The Faculty of the Department of Microbiology, Schools of Medicine and Dentistry State University of New York at Buffalo (Editors), Laboratory Manual for Medical Microbiology. 117 S., 20 Abb., 33 Tab. New York-Basel 1983. Marcel Dekker Inc. SFr. 50.00. ISBN: 0-8247-7040-4.

Dieses Laboratoriumshandbuch ist für Studenten der medizinischen Mikrobiologie geschrieben, dürfte aber auch für das technische Personal in Hospitallaboratorien nützlich sein. Es enthält klare und detaillierte, mit wenig technischem Aufwand ausführbare Prozeduren zur Charakterisierung und Identifizierung der häufigsten menschenpathogenen Mikroorganismen. Die Übungen demonstrieren die wichtigsten Eigenschaften der Organismen und beschreiben diagnostische Prozeduren, die auch die Immunodiagnose von ausgewählten Infektionserregern einschließen. Das Buch eignet sich nicht zum Selbststudium, ohne daß weiterführende Literatur herangezogen wird, da die Ergebnisse und Schlußfolgerungen aus den Übungen fehlen. Es muß auch auf den elementaren Charakter aller Versuche hingewiesen und vor dem Reklametext des Verlages auf dem hinteren Buchdeckel gewarnt werden, der behauptet „. . . this exhaustive collection of illustrated, up-to-date laboratory exercises offers the full range of measures to isolate and identify disease-causing bacteria, fungi, viruses, and protozoa . . .". Dieser Anspruch ist in jeder Hinsicht stark übertrieben.

H. MALKE (Jena)

Zeitschrift für allgemeine Mikrobiologie 24 (1984) 4, 239–245

(Akademie der Wissenschaften der DDR, Forschungszentrum für Molekularbiologie und Medizin Zentralinstitut für Mikrobiologie und experimentelle Therapie, Jena, Direktor: Prof. Dr. rer. nat. habil. U. TAUBENECK)

ATP levels in phosphate-deregulated and arsenate-resistant mutants of *Streptomyces noursei*

F. HÄNEL, U. GRÄFE, W. FRIEDRICH and E. J. BORMANN

(Eingegangen am 26. 9. 1983)

Mycelial levels of ATP and glucose-6-phosphate were investigated in mutants of streptothricin-producing *S. noursei* JA 3890b differing from the wild-type strain in antibiotic formation, in the control by inorganic phosphate of the secondary metabolism, and in the resistance to growth inhibition by toxic arsenate ions. As compared with the ancestral strain, mutants exhibited a lower content of ATP in the mycelium while addition of 0.1 M arsenate to growing cultures provoked only moderate changes in the level of this high-energy metabolite. The results suggest that there exists a correlation between growth resistance to arsenate and insensitivity to phosphate inhibition of the secondary metabolism, on the one hand, and the capacitiy to produce streptothricin-type antibiotics, on the other.

In a recent communication we reported on the selection of arsenate-resistant strains of *Streptomyces noursei* JA 3890b as well as on the biological properties of these clones (FRIEDRICH *et al.* 1984). As it has been shown, strains yielding increased amounts of streptothricin-type antibiotic nourseothricin were discernible from the wild-type strain by either the improved resistance to toxic arsenate ions and/or the relative insensitivity of their antibiotic production to inhibition by excessive concentrations of orthophosphate in the medium. These findings suggested that alterations in the energy metabolism will be a beneficial prerequisite when high-yielding mutants of *S. noursei* should be selected. This conclusion has been supported by ČURDOVÁ *et al.* (1976) and MADRY *et al.* (1979) demonstrating the inverse correlation in different *Streptomyces* strains between the level of two high-energy metabolites, adenosine-5'-triphosphate and glucose-6-phosphate, and the capacity to generate antibiotics, but with regard to the pertinent strains of *S. noursei* JA 3890b no respective data have so far been acquired. Therefore, in the present study we dealt with the levels of ATP and glucose-6-phosphate (G6P) in mutant strains of this microorganism as well as with the effect of arsenate ions on their energy metabolism. The intention was to elucidate conceivable connections between the resistance of growth to arsenate and insensitivity to phosphate inhibition of streptothricin biosynthesis, on the one hand, and the capacity to overproduce streptothricin-type antibiotics, on the other.

Materials and methods

Reagents: Trisodium salt of ATP and glucose-6-phosphate dehydrogenase were purchased from BOEHRINGER (Mannheim, BRD), NADP from VEB Arzneimittelwerke (Dresden, GDR), firefly lantern extract (luciferine-luciferase) and tris buffer from SERVA (Heidelberg, BRD), di-sodium hydrogen arsenate from MERCK (Darmstadt, BRD), carrier-free $^{32}PO_4^{3-}$ from ISOCOMMERZ GMBG (Berlin, GDR), POP and POPOP from REANAL (Budapest, Ungarn), and membrane filters SYNPOR No. 6 (2.5 cm diamter, 0.45 μm pore size) from KAVALIER (CSSR).

Strains and conditions of cultivation: *Streptomyces noursei* JA 3890b (BRADLER and THRUM 1963) was received from the culture collection of the Central Institute of Microbiology and Experimental

Therapy, Jena. Selection and propagation of mutant strains NG 13, UV 12 and UV12-13-As^{R}7 have been described (Friedrich *et al.* 1984). Inocula were grown for two days on Soja 2G medium (g · l^{-1}): glucose 20; soybean meal 20; NaCl 5; KH_2PO_4 0.6; $CaCO_3$ 3; pH 6.7. The main cultivations were carried out at 29 °C on a rotary shaker (240 r.p.m.) in 500 ml shaking flasks containing 80 ml medium (g · l^{-1}): glucose 25; NaCl 5; $CaCO_3$1; NH_4NO_3 3; L-aspartic acid 3; $FeCl_3$ 0.005; $MgCl_2$ · 6 H_2O 0.04; $MnSO_4$ · 2 H_2O 0.02; pH 6.3 (prior to sterilization). Mycelial dry weigth was estimated gravimetrically.

ATP assays: ATP was determined by the bioluminescence method (Strehler 1970, Čurdová *et al.* 1976). Assay mixes contained in 0.8 ml volume: 0.2 ml firefly lantern extract dissolved in 5 ml water and 0.2 ml of each sample. Bioluminescence was measured by the use of a fluorimeter model 8–9 (Locarte, England), and ATP content was calculated from an internal standard of ATP. For adenylate extraction a 10 ml sample was withdrawn from a growing culture, cooled immediately to 4 °C and centrifuged for 2 min at 3000 g. The sediment was boiled for 10 min with 4 ml of distilled water, cooled and then centrifuged for 2 min at 3000 g. The supernatant was frozen and stored at –20 °C for the assay.

Assay of glucose-6-phosphate: Glucose-6-phosphate was estimated according to Hohorst (1972), using the reduction of NADP as a measure. The biomass from 20 ml culture samples was rapidly extracted with 3 ml 6% $HClO_3$ (0 °C) and the mixture centrifuged for 10 min at 10000 g. The supernatant was neutralized with 2 M $KHCO_3$ and after short centrifugation at 15000 g used for the assay. Concentrations in the assay mixture: Tris · HCl buffer 0.2 M, pH 7.6; NADP 0.2 mM; $MgCl_2$ 5 mM; glucose-6-phosphate dehydrogenase 170 mU · ml^{-1} and extract.

Arsenate effect on mycelial ATP concentration: After 48 hrs growth one part of the culture was supplied with 90 mM Na_2HAsO_4 · 7 H_2O. Samples were taken 3 h later for the measurement of intracellular ATP levels.

Uptake of labelled phosphate: The capacity of 24 hrs mycelia to transport $^{32}PO_4^{3-}$ was measured by means of the common technique using membrane filters (Martin *et al.* 1979). 24 hrs cells of both parental and mutant strains were collected, washed two times with 0.1 M KCl and resuspended in the same initial volume of fresh medium differing from the original composition (see above) in that phosphate and $CaCO_3$ have been omitted and concentrations of glucose and $CaCl_2$ were reduced to 12.5 g · l^{-1} and 1 g · l^{-1}, respectively, to make a final concentration of 1–2 mg dry mycelium per ml. After prolonged growth for 5 hours (29 °C) 2 ml samples were transferred in 25 ml shaking flasks, further preincubated for 5 min at 28 °C, supplemented with 10 mM phosphate for 10 min and after that labelled with about 5 · 10^5 cpm · ml^{-1}. At the appropriate time intervals 0.5 ml of each sample were taken from the incubation mixture, filtered rapidly through membrane filters, and the harvested mycelial residue was washed subsequently fourtimes with 1 ml of a 10 mM solution of unlabelled phosphate. The filters were dried and then transferred to scintillation vials containing 2 ml of anhydrous ethanol. 12 ml of scintillation cocktail (5 g PPO; 0.3 g POPOP; 1 l toluene) were added and the measurements were carried out using the LKB-Wallac model 8100 scintillation counter. The values reported were obtained by subtracting blanks prepared in the following manner: an equivalent mycelium sample was treated with 5 · 10^5 cpm · ml^{-1} and without any further incubation immediately filtered through membrane filters and processed as in the aforementioned uptake experiment.

Results

Growth and biological properties of the investigated strains

All strains were grown on a synthetic, soluble, medium which enabled abundant growth and offered the advantage that mycelial biomass could be rapidly sedimented throughout short-time centrifugation. The alternative use of complex, insoluble, media proved unsuccesful for the ATP assays, mainly due to coloured substances which were coextracted with the adenosine-5′-triphosphate giving rise to abnormal quenching of the bioluminescence.

Under the above conditions, the wild-type strain *S. noursei* JA 3890b exhibited similar growth pattern as mutant strains NG 13, UV 12 and UV 12-13-As^{R}7 (Fig. 1a). As found earlier with cultures grown on complex media (Friedrich *et al.* 1984), mutants NG 13 and UV 12 amounted about fivetimes more streptothricins than the parental strain JA 3890b while mutant UV 12-13-As^{R}7 generated an about fourfold higher level of the antibiotic (Fig. 1b). That approximately adequate differences

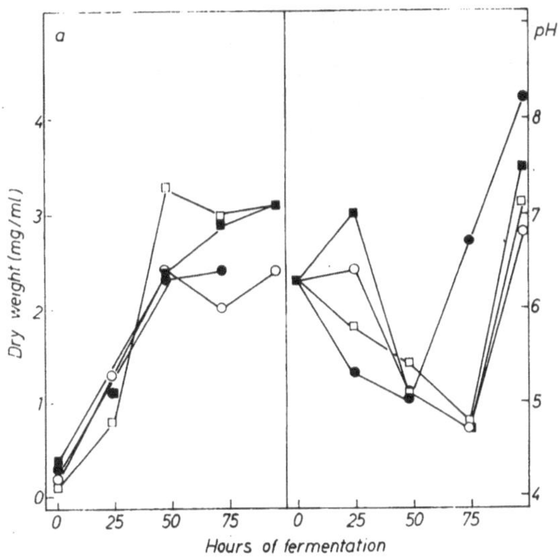

Fig. 1a

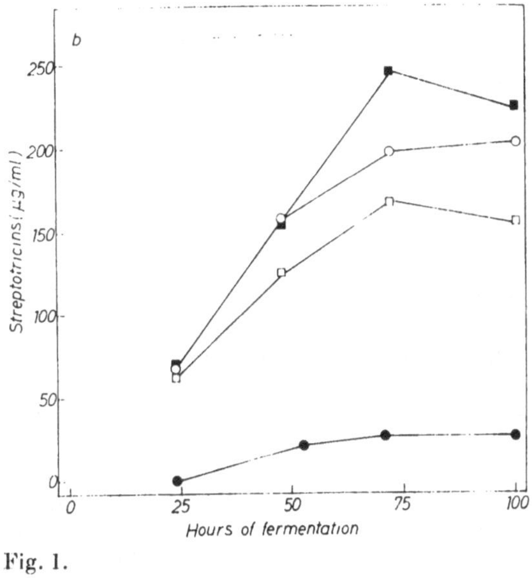

Fig. 1b

Fig. 1.
Growth and pH (1a) and antibiotic production (1b) of *Streptomyces noursei* JA 3890b and his mutant strains during cultivation on a synthetic medium. ● JA 3890b, ○ NG 13, ■ UV 12, □ UV 12-13-As^{R}7

became apparent in the capacity to produce antibiotics in the course of cultivation on both complex and synthetic media suggests that the formation of streptothricins is governed chiefly by genetically determined mechanisms of control.

As far as the inhibition of streptothricin production by excess of inorganic phosphate is concerned, the high-titer strains NG 13 and UV 12 also proved less sensitive than the wild-type strain (FRIEDRICH *et al.* 1984), and selection of the arsenate-resistant strain UV 12-13-As^{R}7 from mutant UV 12 further led to declining sensitivity towards PO_4^{3-} ions. Strain NG 13 already exhibited naturally high resistance to toxic arsenate ions which, noticeably, exceeded that of strain UV 12-13-As^{R}7.

Mycelial levels of ATP and the effect of arsenate ions theoren

Depicted in Fig. 2 is the mycelial level of ATP in the wild-type strain JA 3890b in comparison with the high-producing mutants NG 13, UV 12 and UV 12-13-As^{R}7. The mutants improved in streptothricin biosynthesis displayed significantly reduced

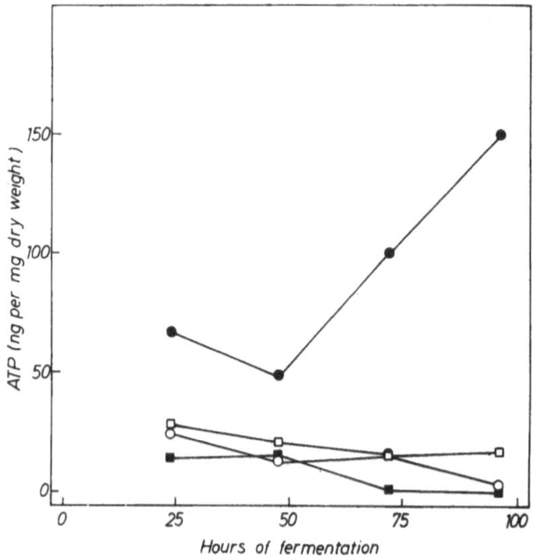

Fig. 2.
Intracellular ATP concentrations of *Streptomyces noursei* JA 3890b and his mutant strains. Symbols as in Fig. 1

levels of ATP during the whole fermentation time indicating that there exists, in fact, an inverse correlation between the generation of streptothricin-type antibiotics and the cellular pool size of this high-energy metabolite. That the second culmination of ATP concentration observed in later stages of cultivation of the parental strain JA 3890b has not been found with the mutants can be interpreted in terms of an altered developmental cycle which does not allow the excessive formation of ATP throughout the stationary phase of cultivation. In contrast to the intracellular level of adenosine-5'-triphosphate, the mycelial pool of glucose-6-phosphate has been less markedly distinguished from that of the wild-type strain JA 3890b in mutants NG 13 and UV 12. But an exception has been the arsenate-resistant strain UV 12-13-As^{R}7 which displayed a comparably lower level of this metabolite in the prolonged course of fermentation (Fig. 3).

As it has been previously shown (FRIEDRICH *et al.* 1984), growth of mutants NG 13 and UV 12-13-As^{R}7 was more resistant to toxic effects of arsenate ions than that of the wild-type strain JA 3890b and mutant UV 12. In order to contribute to better understanding the biochemical background of this feature, we measured the influence of arsenate ions on mycelial pool of ATP in a short-time (3 hrs) experiment. Fig. 4 shows that addition of 0.1 M arsenate provoked subsequent drop by 70% of ATP concentration in the parental strain JA 3890b. At the same time, strain UV 12 appeared as much more sensitive, while mutants NG 13 and UV 12-13-As^{R}7 exhibited only minor changes in their ATP pools, in full accord with the comparably higher resistance to arsenate ions of the latter two strains. It has to be mentioned at this point that under the conditions of cultivation on the synthetic medium the arsenate

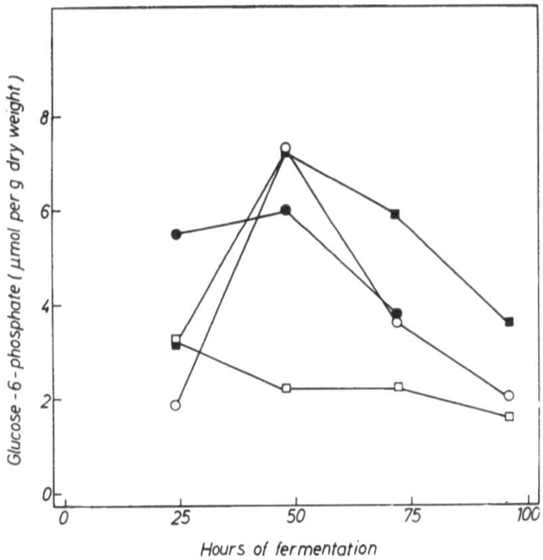

Fig. 3.
Intracellular glucose-6-phosphate concentrations of *Streptomyces noursei* JA 3890b and his mutant strains. Symbols as in Fig. 1

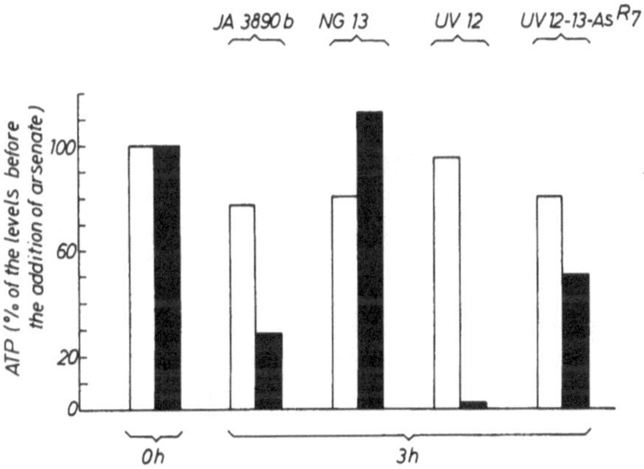

Fig. 4.
ATP levels in *Streptomyces noursei* JA 3890b and his mutant strains grown in the presence (shadowed colums) and absence (open colums) of 90 mM $NA_2HAsO_4 \cdot 7\ H_2O$

resistant strain UV 12-13-As^{R}7 has been proved as a stable phenotype in contrast to earlier findings with complex media (FRIEDRICH *et al.* 1984). Since it could be supposed that the observed phenomenon of resistance to PO_4^{3-} inhibition of antibiotic production could be due to the restricted transport of orthophosphate ions and/or arsenate into the mycelium of strains NG 13 and UV 12-13-As^{R}7, we measured the phosphate uptake by washed 24 hrs mycelia by means of the common membrane filter technique. As shown in Fig. 5 the results of the $^{32}PO_4^{3-}$ incorporation studies reveal that none of the arsenate-restistant strains represented a true permeability mutant, and that, probably, also thet insensitivity to arsenate of ATP pool has to be attributed mainly to changes in the regulation of their energy metabolism.

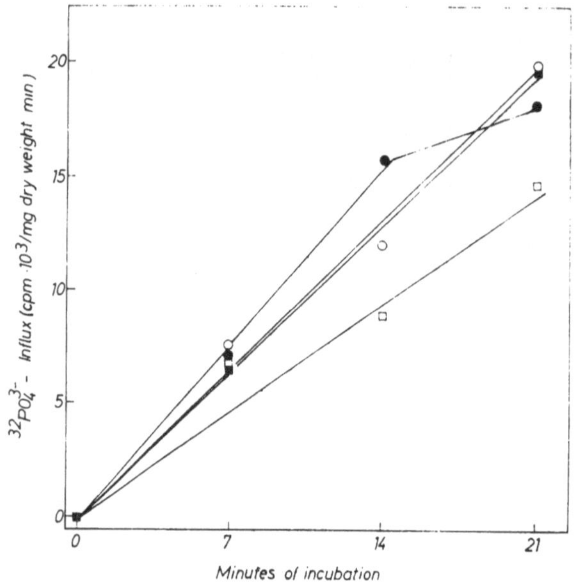

Fig. 5.
Phophate uptake by *Streptomyces noursei* JA 3890b and his mutant strains. Symbols as in Fig. 1

Discussion

Improvement in the production of secondary metabolites should occur as a pleiotropic response to alterations in the expression of very many gene functions and in the activity of many gene products as well. The high-titer mutants of *S. noursei* JA 3890b investigated in this study, for instance NG 13 and UV 12, were selected in the course of a random screening program for high-yielding strains suggesting that their improved productivity may result from several, independent, genetic changes. Thus, mutants UV 12 and NG 13 were discernible from each other and the wild-type strain by their relative resistance to arsenate, a toxic analogue of the orthophosphate ions. Apart from these differences, their cellular ATP pool was drastically reduced as compared with their common ancestor strain, JA 3890b, suggesting that a low rate of energy production represents an essential prerequisite for the installation of a high rate of streptothricin biosynthesis. Obviously, in streptomycetes both metabolic activities, the energy production and the secretion of secondary metabolites, appear to be inversely correlated. This contention is in full agreement with earlier findings of ČURDOVÁ *et al.* (1976) demonstrating reduced ATP levels for the high-yielding strains of chlorotetracycline-producing *S. aureofaciens*. The biochemical consequences of this phenomenon as to the regulation of the overall metabolic machinery yet appear as far from being completely understood since the adenosine-5′-triphosphate and similar high-energy metabolites can be involved in the control of a very large number of metabolic processes. It has been interesting to recognize that the arsenate-resistant mutant UV 12-13-As^{R}7 did not differ from the ancestral strain UV 12 in its cellular ATP level suggesting that changes of the energy metabolism did not provide the biochemical basis for its growth in the presence AsO_4^{3-} ions. Further efforts will be clearly deviced when the reasons for the resistance to arsenate ions should be elucidated more thoroughly. In *Escherichia coli* there are known two genetically separable systems of phosphate transport possessing different substrate affinities, initial velocities, and substrate specificities (WILLSKY and MALAMY 1980). That mutant UV 12-13-As[R]7 was not a permeability mutant is compatible with the view that its resistance to arsenate has to be mainly attributed to the higher specificity of the ATP-regenerating machinery in favour of the selective use of PO_4^{3-} even in the presence of excessive AsO_4^{3-} ions. Additional support has been obtained from the reduced effect of AsO_4^{3-} on mycelial ATP pools in the resistant strains NG 13 and UV 12-13-As[R]7.

Apparently, in the investigated strains there existed a connection between the degree of resistance to arsenate ions and the diminished sensitivity of antibiotic production to inhibition by inorganic phosphate ions. Therefore, selection of arsenate-resistant strains may provide advantages in the further search for improved mutants of this microorganism. But resistance to phosphate inhibition can be assessed to represent only one of many, different, metabolic replenishments which have to be fullfilled in order to stimulate antibiotic production. It is, probably, due to this reason, that the majority of the arsenate-resistant strains of *S. noursei* failed to yield increased titres of streptothricins (FRIEDRICH *et al.* 1984). Recent communications have outlined the possible role of glucose-6-phosphate as the presumed mediator of carbon catabolite repression of antibiotic biosynthesis (MADRY *et al.* 1979). In contrast to these observations, however, in the mutants of *S. noursei* JA 3890b no noticeable differences could be detected in the cellular pools of this metabolite. But this finding can be readily assessed when one takes into consideration that streptothricin biosynthesis by *S. noursei* is not subjected to catabolite repression by excessive glucose in the medium.

Acknowledgements

The authors gratefully acknowledge the technical assistance of Miss G. KÖRBS and Mrs. CH. WEISER.

References

BRADLER, G. und THRUM, H., 1963. Nourseothricin A und B, zwei neue antibakterielle Antibiotika einer *Streptomyces noursei*-Variante. Z. allg. Mikrobiol., **3**, 105–112.

ČURDOVÁ, E., KŘEMEN, A., VANĚK, Z. and HOŠŤÁLEK, Z., 1976. Regulation of biosynthesis of secondary metabolites. XVIII. Adenylate level and chlorotetracycline production in *Streptomyces aureofaciens*. Folia microbiol., **21**, 481–487.

FRIEDRICH, W., BORMANN, E. J. and GRÄFE, U., 1984. Isolation and biological properties of arsenate-resistant strains of *Streptomyces noursei*. Z. allg. Mikrobiol., **24**, 13–19.

HOHORST, H.-J., 1970. D-Glucose-6-phosphat and D-Fructose-6-phosphat. In: Methoden der enzymatischen Analyse Vol. II (Editor: H. U. BERGMEYER). Akademie Verlag Berlin, S. 1200 bis 1204.

MADRY, N., SPRINKMEYER, R. and PAPE, H., 1979. Regulation of tylosin synthesis in *Streptomyces*: effects of glucose and inorganic phosphate. Europ. J. Appl. Microbiol. Biotechnol., **7**, 365–370.

MARTIN, J. F., NAHARRO, G., LIRAS, P. and VILLANUEVA, J. R., 1979. Isolation of mutants deregulated in phosphate control of candicidin biosynthesis. J. Antib., **32**, 600–606.

STREHLER, B. L., 1970. Adenosin-5-triphosphat und Creatinphosphat. Bestimmung mit Luciferase. In: Methoden der enzymatischen Analyse Vol. III (Editor: H. U. BERGMEYER). Akademie Verlag Berlin, S. 2036–2050.

WILLSKY, G. R. and MALAMY, M. H., 1980. Characterization of two genetically separable inorganic phosphate transport systems in *Echerichia coli*. J. Bacteriol., **144**, 356–365.

Mailing address: Dr. F. HÄNEL, Zentralinstitut für Mikrobiologie und experimentelle Therapie der AdW, DDR-6900 Jena, Beutenbergstraße 11

Buchbesprechungen

A. FIECHTER (Editor), Reaction Engineering. 171 S., 99 Abb., 44 Tab. Berlin 1983. Akademie-Verlag. M 82,00.

Das vorliegende Buch stellt den Band 24 der Reihe „Advances in Biochemical Engineering" dar. Es enthält vier unabhängige Review-Artikel von großer Praxisnähe, die auf moderne Entwicklungen der einzelnen Gebiete eingehen und Grundlagen sowie den Stand darstellen.

Die erste Arbeit befaßt sich in ausführlicher Darstellung mit dem Problem der Gaslöslichkeit in Kulturlösungen. Es wird eine einfache logarithmisch-additive Annäherungsmethode vorgestellt, welche unter Einbeziehung von Löslichkeitsparametern (K_1) der einzelnen Nährmedienkomponenten ein Abschätzen der Sauerstofflöslichkeit ermöglicht. Für eine Vielzahl von Medienkomponenten sind die Löslichkeitsparameter aufgelistet. Fehlermöglichkeiten werden diskutiert. Die zweite Arbeit beinhaltet Probleme und Auslegungsmöglichkeiten von mikrobiellen Biokatalysatoren bei der Anwendung im Industriemaßstab. Es wird auf Möglichkeiten für ein optimales Katalysatordesign eingegangen und die dafür bedeutsamen Parameter analysiert sowie über Entwicklungen und Einsatzbereiche berichtet. Die beiden letzten Arbeiten gehen auf das Gebiet der Stofftrennung ein, wobei die Flüssig-Flüssig-Extraktion bei der Proteinreinigung und die Fest-Flüssig-Separation zur Biomasseabtrennung behandelt werden. Es werden insbesondere Verfahren im Produktionsmaßstab und Möglichkeiten zur Prozeßoptimierung von Verfahren mit hoher Selektivität dargestellt. Auf Berechnungsgrundlagen von Abtrennraten wird ebenfalls eingegangen.

M. MENNER (Jena)

W. M. FOGARTY (Editor), Microbial Enzymes and Biotechnology. XI + 382 S., 70 Abb., 73 Tab. London - New York 1983. Applied Science Publishers. Ł 34.00. ISBN 0-85334-185-0.

Namhafte, auf dem Gebiet der technischen Enzymologie tätige Autoren berichten über neuere Ergebnisse zur Produktion, Charakteristik und Anwendung einiger praxisrelevanter mikrobieller Enzyme. Besondere Berücksichtigung findet der Lebensmittelsektor, wenngleich auch andere Applikationsbereiche behandelt werden (z. B. Verwertung cellulosehaltiger Ab- und Nebenprodukte, Waschmittelindustrie, medizinischer Bereich u. a.). Folgende Enzymgruppen werden erörtert: Amylasen (lösliche und trägerfixierte α-Amylase, Amyloglucosidase, β-Amylase, entzweigende Enzyme, α-Glucosidase, Schardinger-Dextrin-synthetisierende und -abbauende Enzyme); Glucoseisomerase, Glucoseoxydase; pektinolytische Enzyme (Esterasen, Hydrolasen, Lyasen); Cellulasen (Endoglucanasen, Cellobiohydrolase, β-Glucosidase); Lipasen; Proteinasen (Peptidasen; Serin-, Thiol-, Metall-, saure Proteasen). Ein Kapitel über die bei der mikrobiellen Enzymsynthese ablaufenden regulatorischen Prozesse beschließt die Ausführungen.

Sämtliche Kapitel sind mit einem aktuellen Literaturverzeichnis ausgestattet, so daß dem interessierten Leser ein rasches Einarbeiten in Spezialgebiete ermöglicht wird. Die einzelnen Kapitel sind hinsichtlich der Untergliederung allerdings nicht einheitlich gestaltet, die Akzente von den Autoren jeweils unterschiedlich gesetzt. Die verwendete Enzym-Nomenklatur entspricht nicht immer den neuesten Empfehlungen des Nomenklatur-Komitees der IUB. So wird z. B. im Falle der pektinolytischen Enzyme die Fassung von 1973 (statt von 1978) herangezogen.

H. RUTTLOFF (Potsdam)

Zeitschrift für allgemeine Mikrobiologie 24 (1984) 4, 247–252

(Department of Agricultural Biology, University of Ibadan, Ibadan, Nigeria)

Isolation, purification and assay of the macerating enzyme produced by *Penicillium oxalicum* CURIE and THOM

T. IKOTUN

(Eingegangen am 23. 9. 1983)

Penicillium oxalicum produced two isozymes of polygalacturonase (PG) and a pectate lyase (PL). The enzymes were separated and purified following ammonium sulphate precipitation, ion exchange chromatography, ultrogel column chromatography and isoelectric focusing. The first isozyme of polygalacturonase (PGI) was rather unstable hence its properties could not be much assayed. PGII macerated and killed yam tissue in 4 hours but PL was unable to do so. Enzyme assay for the end-products of degradation of sodium polypectate and yam tissue showed that PGI was an exo-enzyme while PGII and PL were endo-enzymes. Endo-polygalacturonase (PGII) appears to play the major role (as the macerating enzyme) in the pathogenesis of yam tissue infected by *P. oxalicum*.

Pectic enzymes produced by phytopathogenic organisms have often been associated with maceration of plant tissue (BATEMAN 1968, BATEMAN and MILLAR 1966, SATO 1968). In most cases, endo-pectolytic enzymes (hydrolases and lyases) which degrade in a random manner, the a-1, 4 linkages between galacturonsyl moieties in polymers of galacturonic acid appear to be the primary agents responsible for the maceration process (BATEMAN 1968, BYRDE and FIELDING 1968, MOUNT *et al.* 1970, SATO 1968, STARR and MORAN 1962). In many cases, the macerating enzyme was identified as an endo-polygalacturonate *trans*-eliminase (BASHAM and BATEMAN 1975, MOUNT *et al.* 1970), and in some cases, it was an endo-polygalacturonase (BATEMAN 1968, 1972). An endo-pectinmethyl *trans*-eliminase has been identified as the macerating enzyme produced by *Sclerotinia fructigena* (BYRDE and FIELDING 1968). The other cell wall-degrading enzymes, namely, cellulases, hemicellulases *etc.* have not been shown to macerate plant tissue (BAUER *et al.* 1977).

One of the characteristics of brown rot of yam tubers caused by *Penicillium oxalicum*, is the destruction of invaded tissue. This is often soft, depending on the amount of moisture present in the immediate environment (ADENIJI 1970), and whether a bacterium invades the infected tissue.

These observations and the fact that *P. oxalicum* spreads through yam tissue mainly intercellularly suggest that pectolytic enzymes are involved in pathogenesis. The softness of the invaded tissue also suggest that it may have been macerated by the fungus. This work was, therefore, designed to isolate and purify the macerating enzyme of *P. oxalicum* and to study its effects on yam tissue.

Materials and methods

P. oxalicum was isolated from infected yam (*Dioscorea rotundata*) tissue and grown on potato dextrose agar (PDA) at 27 °C. After pure growth was established, an 8 mm diameter agar disc containing fungal mycelia and spores was cut out with the aid of a sterile size 4 corkborer. This was used to inoculate autoclaved (at 1.1 kg/cm^2 for 20 min) and cooled liquid medium of BAUER *et al.* (1977) which was supplemented with 0.5 g sodium polypectate (NaPP)/litre in 2.8 litre FERNBACH flasks.

Inoculated flasks were incubated at 27 °C for 8 days after which the content was filtered through an 8-layer cheese cloth and centrifuged at 5,500 × g for 30 min. Culture filtrate was then subjected to fractionation at 40, 60, 80 and 95% ammonium sulphate saturation. Precipitate from each fraction was separately taken up in 20 ml of 100 mM citrate-phosphate buffer for enzyme assays at pH 3.0—5.0, 100 mM phosphate buffer at pH 6.0—7.0 and 50 mM tris-HCl buffer at pH 8.0—9.0. These were dialysed for at least 15 hours against respective buffers at 4 °C and each fraction was assayed for polygalacturonase and pectate lyase activities at pH values reported above. Fractions containing the highest enzyme activity were then purified.

Since 80 and 95% fractions from ammonium sulphate precipitation contained the highest pectolytic activity (Fig. 1), they were pooled together and subjected to diethyl-aminoethyl cellulose (DEAE-Cl form) ion exchange chromatography in a 1.5 × 15 cm column at 4 °C. The column was equilibrated with 50 mM tris-HCl buffer (pH 8.0).

About 50 ml of the pooled fractions were applied to the column and were washed down with 30 ml of the same buffer and then eluted with buffer-500 mM NaCl gradient. Fractions of 5 ml were collected and assayed for enzyme activity. Peaks of activity were then subjected to AcA54 Ultrogel column chromatography at 4 °C and eluted with 50 mM phosphate buffer, pH 7.0. Again, fractions were collected in 5 ml portions in an LKB fraction collector.

Peaks of pectolytic enzyme activity from the ultrogel column were exhaustively dialysed against distilled water at 4 °C and then subjected to electrofocusing a in granulated bed of Ultrodex using an LKB Multiphor apparatus (Model 2117, LKB Instuments Inc. Rockville MD 20852). The methods suggested by the manufacturers (LKB application note 198) were followed using pH 3—10 Ampholine carriers (2%). After 15 hours, the gel bed (10.7 × 23 cm) was divided into 30 equal zones using a ready-made grid (an accessory). The gel from each zone was scooped out and placed into a 1.5 × 5 cm column and eluted with 3 ml of distilled water and the pH of the eluate was measured. Each eluate was then assayed for polygalacturonase and pectate lyase activities. The isoelectric point (pI) of the peak of each enzyme activity was noted. Purified enzymes so obtained from all the above apparatus were again assayed from pH 3.0—9.0 to detect the optimum pH of activity.

Polygalacturonase activity was assayed viscometrically and by the increase in free reducing groups in reaction mixtures according to the methods of NELSON (1944). Enzyme activity was expressed as μmoles of unsaturated galacturonide released/ml of enzyme x min. Lytic cleavage of NaPP was determined by following the increase in absorbance of reaction mixture at 230 nm (STARR and MORAN 1962) and by the periodate-thiobarbituric acid (TBA) method (WEISSBACH and HURWITZ 1959). For both assays, one unit of enzyme activity was taken as that amount which produced 1 μmole of unsaturated galacturonide/min.

The effect of the macerating enzyme on yam tissue was studied by the Neutral Red assay (BASHAM and BATEMAN 1975). Tissue maceration and cell death were the two effects studied. A cell death index (BASHAM and BATEMAN 1975) of 0—5 on a linear scale on which 0 indicated that all cells accumulated the dye (all living) and 5 indicated that no cell accumulated the dye (all dead) was used. Tissue maceration was observed on yam discs stained with Neutral Red dye.

The degree of cohesion between cells in the discs was determined by estimating the ease with which the discs could be pulled apart with mounted needles or by the amount of force required to press through the discs using a pair of forceps. Maceration index was based on the linear scale of 0—5, on which 0 indicated cellular cohesion (no maceration) and 5 indicated no cohesion (tissue was macerated). In control experiments, boiled enzymes were used in place of active enzymes.

Assay for the end-products of degradation of sodium polypectate (NaPP) and of yam cell wall by the pectolytic enzymes was carried out by paper chromatography. A 0.5% (W/V) solution of NaPP was prepared in 100 mM citrate-phosphate buffer at pH 5.0 and in 50 mM tris-HCl buffer at pH 8.0. One ml of a solution of NaPP in appropriate buffers were incubated with 1.0 ml of enzymes separated and isolated by the various purification processes at 30 °C in a water bath. At intervals of 30 min up to 4 hours and later at intervals of 4 hours up to 12 hours, 50 μl of the well agitated reaction mixture were removed in a tube and boiled for 10 min in a water bath to stop enzymatic reaction.

Results and discussion

Fig. 1 shows that 80 and 95% fractions from ammonium sulphate precipitation contained the highest pectolytic enzyme activity. It also shows peaks of enzyme activity at pH 5.0 and 8.0. Both fractions were pooled together. After passage through the ion exchange column, samples were collected in 5 ml portions from an LKB

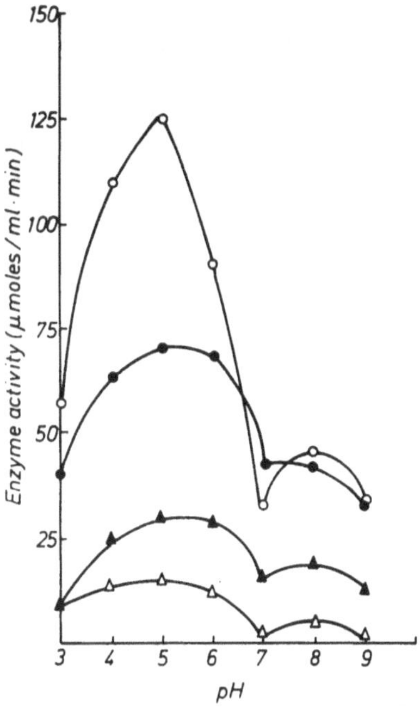

Fig. 1.
Effect of pH on pectolytic enzyme activity of different fractions of ammonium sulphate saturation. Culture filtrate (1 litre) was concentrated down by this method. ○ 95%, ● 80%, ▲ 60%, △ 40% ammonium sulphate saturation

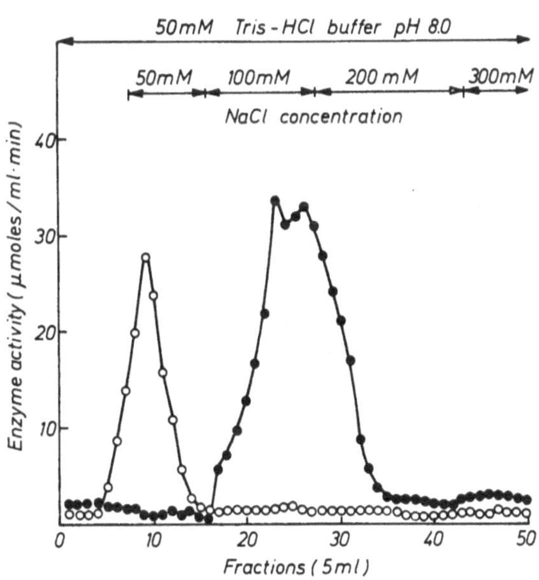

Fig. 2.
Separation of pectate lyase (●) and polygalacturonase (○) from filtrate of culture of *Penicillium oxalicum*

fraction collector and assayed for enzyme activity. The profile of pectolytic enzymes separated by this procedure is given in Fig. 2. The peak of pectate lyase activity was collected at fraction number 9 while polygalacturonase had two peaks at numbers 23 and 26. Peaks of enzyme activities were passed through an ACA 54 ultrogel column at 4 °C and eluted with 50 mM phosphate buffer, pH 7.0. The profile of enzymes from the ultrogel column given in Fig. 3 shows a good separation of the pectate lyase (PL)

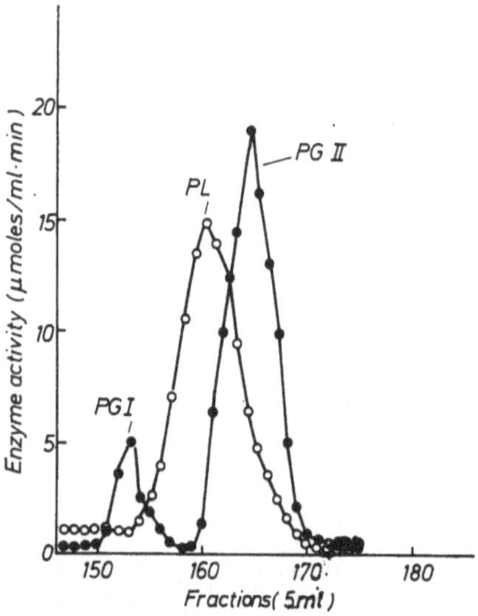

Fig. 3.
Profile of separated enzymes from the ultrogel AcA 54 column

from the polygalacturonase. Two isozymes of polygalacturonase were obtained. The first (PGI) had its peak in fraction number 153 and the other (PGII) had its peak of activity in fraction number 164. The peak of activity of pectate lyase was found in fraction number 160.

Fig. 4 shows the effect of pH on the activity of PGII and PL. At pH 5.0, PGII was the more active while PL activity was suppressed. At pH 8.0, PL activity was high while PGII activity was suppressed.

Only two enzymes were recovered from the flat bed electrofocusing apparatus, PL and PGII (Fig. 5). The pectate lyase had its peak in fraction number 16 and its isoelectric point (pI) at pH 4.9. PGII was recovered in fraction number 24 and had its pI at pH 3.6. Polygalacturonase I was not recovered from the apparatus.

Peaks of activity of PGII and PL obtained from the ultrogel column were used in the assay for cell death and tissue maceration. Results (not presented) indicate that yam tissue was not macerated by PL. PGI was too unstable to be used in this assay. Only PGII macerated yam tissue. Maceration started within 30 min and by 4 hours, the tissue had been macerated (loss of coherence, scale 5). Incipient cell death was noticed within 4 hours after exposure of yam tissue to PGII. Total loss of ability to accumulate Neutral Red Stain was observed 24 hours after exposure of the tissue to PGII.

PL did not cause maximum cell death. Incipient cell death was not noticed within 2 hours but even 24 hours after, maximum cell death was not realised. It is, therefore, apparent that PGII is the macerating enzyme and that tissue maceration (cell

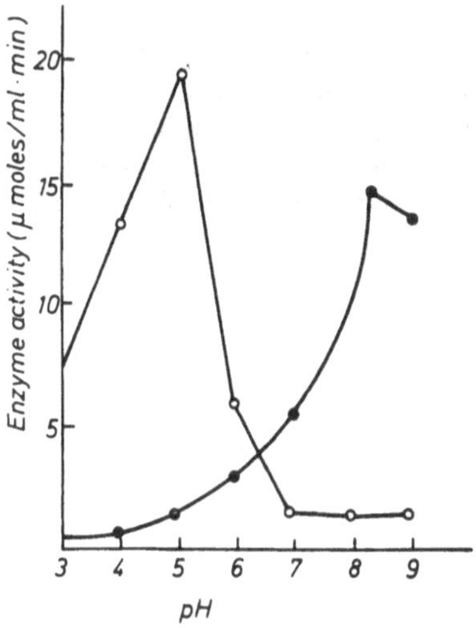

Fig. 4.
The pH optima of activity of polygalacturonase II and pectate lyase separated from the ultrogel AcA 54 column. PG II (○), PL (●)

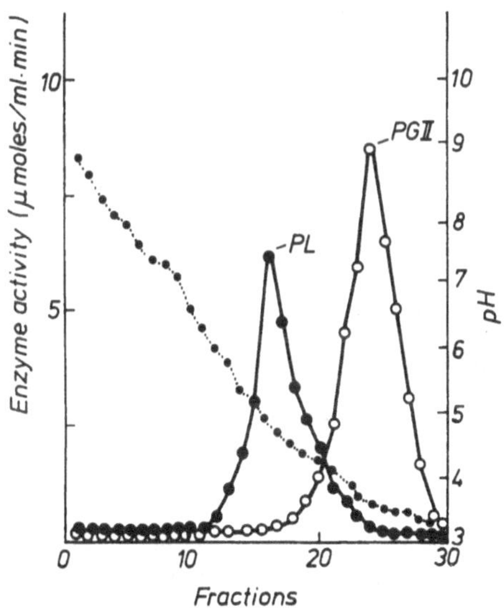

Fig. 5.
Enzyme profile obtained from an LKB Multiphor electrofocusing apparatus using pH 3—10 Ampholine carriers (2%). pH values of isoelectric points of peaks of enzyme activity (··············)

separation) preceded cell death. The pH of optimum activity of PGII when compared with the low pH of yam tissue infected by *P. oxalicum* indicate that PGII probably plays the major role in pathogenesis. The pH of optimum activity of pL (pH 8.0) is too high for this enzyme to play any major role in pathogenesis.

Assay for the end products of degradation of NaPP and yam cell walls by the enzymes show that PGI released only monogalacturonides and hence is an exopolygalacturonase. PGII and PL both released oligogalacturonides, hence are endo-enzymes. This classification is in consonance with earlier work (Bateman 1968, Mount *et al.* 1970, Sato 1968), which show that only random cleavage of the a-1, 4 linkages between galacturonsyl moieties could release oligogalacturonides from cell walls, and, enzymes implicated were the endo-types. These have been responsible for the maceration of plant cell walls (Bateman 1968, 1972, Bateman and Millar 1966, Sato 1968).

Acknowledgements

The author is grateful to the United States Council for International Exchange of Scholars (CIES) Washington D.C., for an award of a Hays-Fulbrigth Fellowship, through which this work was possible. The author also thanks the Department of Plant Pathology, Cornell University, Ithaca, NY 14853, U.S.A. and its entire staff for providing working facilities and for being so helpful and friendly.

References

Adeniji, M. O., 1970. Influence of moisture and temperature on yam decay microorganisms. Phytopathology, **60**, 1968—1969.

Basham, H. G. and Bateman, D. F., 1975. Relationship of cell death in plant tissue treated with a homogeneous endo-pectate lyase to cell wall degradation. Physiol. Plant Path., **5**, 249—262.

Bateman, D. F., 1968. The enzymatic maceration of plant tissue. Nether. J. Plant Path., **74**, 67—80.

Bateman, D. F., 1972. The polygalacturonase complex of *Sclerotium rolfsii*. Physiol. Plant Path., **2**, 175—184.

Bateman, D. F. and Millar, R. L., 1966. Pectic enzymes in tissue degradation. Ann. Rev. Phytopathol., **4**, 119—146.

Bauer, W. D., Bateman, D. F. and Whalen, C. H., 1977. Purification of an endo-a-1, 4-galactanase produced by *Sclerotinia sclerotiorum*: Effect on isolated plant cell walls and potato tissue. Phytopathology, **67**, 867—868.

Byrde, R. J. W. and Fielding, A. H. 1968. Pectinmethyltrans-eliminase, the macerating factor of *Sclerotinia fructigena* and its significan cein brown rot of apple. J. Gen. Microbiol., **52**, 287—297.

Garibaldi, A. and Bateman, D. F., 1971. Pectic enzymes produced by *Erwinia chrysanthemi* and their effects on plant tissue. Physiol. Plant Pathol., **1**, 25—40.

Lowry, O. H., Rosebrough, N. J., Farr, A. L. and Randall, R. J., 1951. Protein measurement with the Folin-Phenol reagent. J. biol. Chem., **193**, 265—275.

Mount, M. S., Bateman, D. F. and Basham, H. G., 1970. Induction of electrolyte loss, tissue maceration and cellular death of potato tissue by an endopolygalacturonate *trans*-eliminase. Phytopathology, **60**, 924—931.

Nelson, N., 1944. A photometric adaptation of the Somogyi method for the determination of glucose. J. biol. Chem., **153**, 375—380.

Sato, S., 1968. Enzymatic maceration of plant tissue. Physiol. Plant., **21**, 1067—1075.

Starr, M. P. and Moran, F., 1962. Eliminative split of pectic substances by phytopathogenic soft rot bacteria. Science, **135**, 920—921.

Weissbach, A. and Hurwitz, J., 1959. The formatin formation of 2-keto-3-deoxy heptonic acid in extracts of *Escherichia coli* B. J. biol. Chem., **234**, 705—709.

Mailing address: Dr. T. Ikotun, Department of Agricultural Biology, University of Ibadan, Ibadan, Nigeria

Zeitschrift für allgemeine Mikrobiologie 24 (1984) 4, 253–259

(Zentralinstitut für Mikrobiologie und experimentelle Therapie der Akademie der Wissenschaften der DDR, Jena[1]), und Institut für Biochemie und Physiologie der Mikroorganismen der Akademie der Wissenschaften der UdSSR, Puschtschino[2])

Incorporation of $^{18}O_2$ during cometabolic degradation of 3-chloroaniline by *Rhodococcus* sp. An 117

D. JANKE[1]), B. P. BASKUNOV[2]), M. YU. NEFEDOVA[2]), A. M. ZYAKUN[2]) and L. A. GOLOVLEVA[2])

(Eingegangen am 28. 9. 1983)

Studies on the effect of different oxygen supply on cometabolic degradation of 3-chloroaniline by resting aniline-grown cells of *Rhodococcus* sp. An 117 revealed that atmospheric oxygen is required for the enzymatic reaction initiating this degradative process which under "normal" conditions (*i.e.* cell incubation in air atmosphere) is characterized by a transient accumulation of 4-chlorocatechol. In oxygen atmosphere, however, no appreciable amounts of 4-chlorocatechol were detected in the incubation medium, probably because of an activation of the ring-cleavage reaction in An 117 cells. Instead of, a nonchlorinated compound accumulated which was identified as the γ-lactone of β-hydroxymuconic acid (γ-carboxymethylene-$\Delta^{\alpha\beta}$-butenolide). The results of experiments with $^{18}O_2$ as well as $H_2{}^{18}O$ indicate that formation of the latter intermediate was due to incorporation of four atoms of atmospheric oxygen and that the initial conversion of monochlorinated anilines into chlorocatechols by *Rhodococcus* sp. An 117 is a dioxygenase-mediated reaction.

Chlorinated anilines are well-established intermediates in degradation of various phenylurea, acylanilide, and phenylcarbamate herbicides. With a few exceptions (ZEYER and KEARNEY 1982), they do not support microbial growth, but are subjected to cometabolic degradation by certain aniline-degrading bacterial cultures supplied with suitable additional carbon sources (REBER *et al.* 1979, SUROVTSEVA *et al.* 1980, YOU and BARTHA 1982). This is also true of aniline-grown cells of *Rhodococcus* sp. An 117 which degrade 2- and 3-chloroaniline with rates of 43 and 57% (compared with the unsubstituted aniline) when being incubated in phosphate buffer containing glucose as the additional carbon source (SCHUKAT *et al.* 1983). The kinetics of this cometabolic process were characterized by a transient accumulation of 3- and 4-chlorocatechol, respectively. Furthermore, evidence was obtained that the stimulating effect of glucose was mainly associated with conversion of 2- and 3-chloroaniline into the respective chlorinated catechols (SCHUKAT *et al.* 1983). By conducting experiments with oxygen-18 we now studied the nature of this enzymatic reaction initiating monochloroaniline degradation in *Rhodococcus* sp. An 117. The results obtained are consistent with formation of the chlorocatechols due to the action of a deaminating dioxygenase and additionally allow some insights into the pathway used by *Rhodococcus* sp. An 117 for cometabolic degradation of 2- and 3-chloroaniline.

Materials and methods

Isolation and identification of *Rhodococcus* sp. An 117, the growth conditions used, and the techniques for estimation of anilines were described previously (KAMINSKI *et al.* 1983). Metabolites were extracted from the incubation media with diethyl ether (SKRYABIN *et al.* 1976). Thin-layer chromatography was performed by using silica-gel plates with fluorescent indicator (KAVALIER, CSSR) and the solvent system benzene: ethanol (8:2, vol/vol). The fractions were located by means of ultraviolet light, eluated with methanol and subsequently applied to chromato-mass-spectrometry.

The experiments in nitrogen or oxygen atmosphere were performed by shaking 10-ml-suspensions of aniline-grown An 117 cells in two-necked-flasks (50 to 60 cm³) at 28 °C. After addition of the substrates the flasks were alternatively evacuated and flushed with either N_2 or O_2 three times. In the experiments with oxygen-18 gas a vial containing 100 cm³ of $^{18}O_2$ (about 95%) was connected with the reaction flask which was previously three times evacuated and flushed with nitrogen.

Mass spectra were obtained with a LKB-2901 type gas chromatographmass spectrometer (ionizing current, 70 e.V.).

^{1}H-NMR spectra were recorded by means of a VARIAN HA-100D type spectrometer (USA), with tetramethylsilane used as internal standard.

Infra-red spectra were recorded in KBr with a type UR-20 spectrograph (VEB CARL-ZEISS Jena, GDR).

The melting point of compound X was determined by using a BOETIUS melting-point apparatus.

Results

During cometabolic degradation of 2- and 3-chloroaniline by resting aniline-grown cells of *Rhodococcus* sp. An 117 a transient accumulation of *o*-dihydroxybenzenes was observed (SCHUKAT *et al.* 1983). A reidentification of these metabolites by means of chromato-mass spectrometry confirmed their identities with 3- and 4-chlorocatechol, respectively. Besides, in the case of 2-chloroaniline small amounts of a second metabolite were detected which was identified as the 2-chloromuconic acid by the same method.

This was the starting point for studies on incorporation of oxygen-18 from $^{18}O_2$ and $H_2{}^{18}O$ during 3-chloroaniline turnover by aniline-grown An 117 cells which were undertaken in order to obtain information on the nature of the initial conversion of anilines by that strain. However, preliminary experiments on cometabolic degradation of 3-chloroaniline under conditions of different oxygen supply revealed that change of air by oxygen within the reaction flasks resulted in the accumulation of a compound not reacting when sprayed with 3% $FeCl_3$ although its R_f-value (0.38) was very similar to that of 4-chlorocatechol (0.35). The same compound accumulated when the reaction was carried out in phosphate buffer enriched with $H_2{}^{18}O$ (about 70%) (Table 1).

Table 1

Relative rates and accumulating metabolites of turnover of 2-chloroaniline (2-CA) and 3-chloroaniline (3-CA) by aniline-grown cells of *Rhodococcus* sp. An 117 in 0.1 M phosphate buffer[a]), pH 6.9, under different conditions of oxygen supply

substrate	atmosphere	medium	relative rate (in %)	metabolites accumulated
2-CA	air	H_2O	76	3-chlorocatechol 2-chloromuconic acid
3-CA	air	H_2O	100	4-chlorocatechol
	N_2	H_2O	3	
	$^{18}O_2$	H_2O	130	metabolite X
	air	$H_2{}^{18}O$	92	metabolite X

a) initial concentrations: monochloroanilines, 1 mM; biomass, 0.6–1.0 mg dry weight of cells per milliliter; glucose, 500 mg/l

Identification of the unknown metabolite accumulating during cometabolic 3-chloroaniline degradation in oxygen atmosphere

To obtain the unknown metabolite in amounts sufficient for physico-chemical analysis, two 250-ml-suspensions of aniline-grown An 117 cells (initial biomass conc., 1.6 mg dry weight of cells per milliliter of 0.1 M phosphate buffer, pH 6.9) were supplied with 3-chloroaniline (1 mM) and glucose (2.78 mM), alternatively evacuated and flushed with $^{18}O_2$ three times and shaken at 28 °C for 2.5 hours. The supernatants obtained after removal of the cells by centrifugation were treated as described above (Materials and methods) except that in this case the resulting ether extract was applied to preparative thin-layer chromatography on MERCK silica-gel plates (Kieselgel 60 F_{254}, 1 mm, 20 cm × 20 cm). The compound in question was then eluted with diethyl ether. Evaporation of the solvent revealed white crystals (total yield, 14 mg; melting point, 164–166 °C; λ_{max} in CH_3OH – 277 nm) which were subsequently subjected to physico-chemical analysis.

From the mass spectrum (Fig. 1a) showing a parent peak at m/e 140 it could be concluded that the isolated compound did not contain organically bound chlorine. The IR-spectrum (Fig. 2) showed absorption peaks at 1780, 1680, 1560 and 1630 cm^{-1}

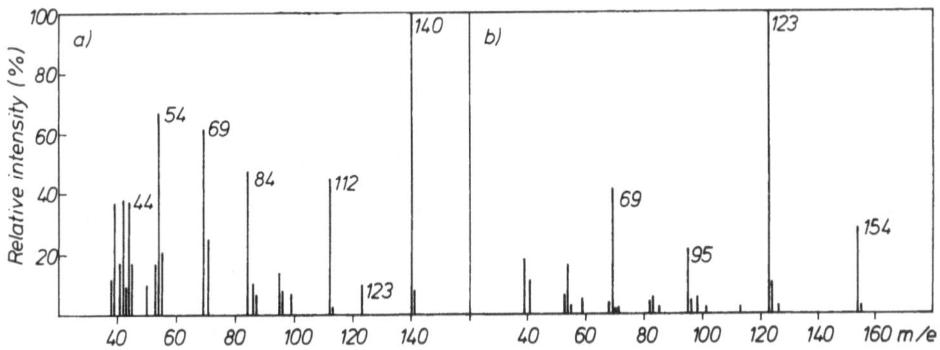

Fig. 1.
Mass spectra of (a) metabolite X accumulating during cometabolic 3-chloroaniline turnover by resting aniline-grown An 117 cells in $^{16}O_2$-atmosphere, and (b) the respective methyl ether

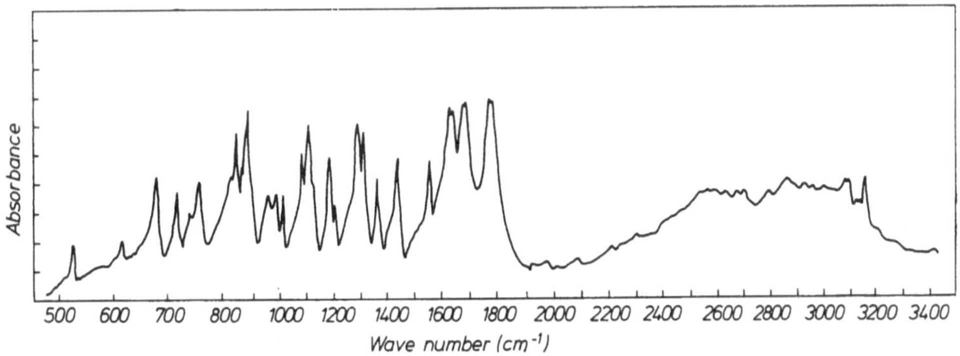

Fig. 2.
Infra-red spectrum of metabolite X accumulating during cometabolic 3-chloroaniline turnover by resting aniline-grown An 117 cells in $^{16}O_2$-atmosphere

indicating the presence of a five-membered lactone, a C = O-group, and of one internal and one external double bond (–C = C–), respectively. The absorption observed at 2500–2700 cm^{-1} corresponds to the presence of a carboxylic group associated with a double bond (dimer formation). On the basis of these data it was supposed that the metabolite in question is identical with the γ-lactone of β-hydroxymuconic acid

(γ-carboxymethylene-$\Delta^{\alpha\beta}$-butenolide) isolated by EVANS *et al.* (1971) from cultures of a soil pseudomonad grown with 4-chlorophenoxyacetate as sole source of organic carbon. The proton magnetic resonance spectrum of the metabolite in D-methanol revealed three single proton-signals which are fully consistent with the structure (Fig. 3) proposed.

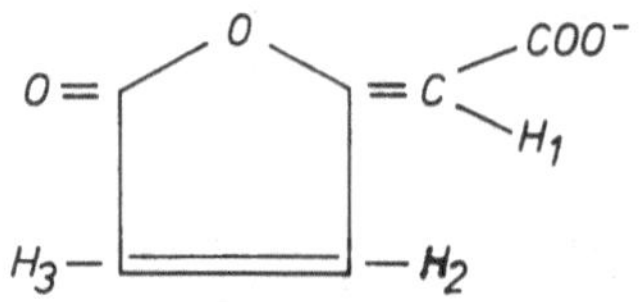

Fig. 3.
^{1}H-NMR data of metabolite X (in D-methanol). $\delta_{H_1} = 5.90$ ppm, $\delta_{H_2} = 6.57$ ppm, $\delta_{H_3} = 8.39$ ppm; coupling constants: $J_{H_1H_2} = 2.0$ Hz, $J_{H_1H_3} = 0.7$ Hz, $J_{H_2,H_3} = 6.0$ Hz

Incorporation of oxygen-18

In nitrogen atmosphere there was practically no turnover of 3-chloroaniline (Table 1) demonstrating that the reaction initiating the cometabolic process required the presence of atmospheric oxygen. Therefore it was reasonable to attribute the formation of chlorocatechols from 2- and 3-chloroaniline in the control variants (air atmosphere) to an oxygenative conversion of the aromatic substrates. To prove this, experiments with $^{18}O_2$ and $H_2{}^{18}O$ were performed to investigate the incorporation of oxygen-18 into the γ-lactone of β-hydroxymuconic acid accumulating during cometabolic conversion of 3-chloroaniline by aniline-grown An 117 cells under these conditions (Table 1).

As shown in Fig. 4 (B) mass analysis of the product accumulating in $^{16}O_2$-atmosphere revealed a spectrum with a parent peak at m/e 140. A quite analogous spectrum was obtained for the compound accumulating in phosphate buffer enriched (about 70%)

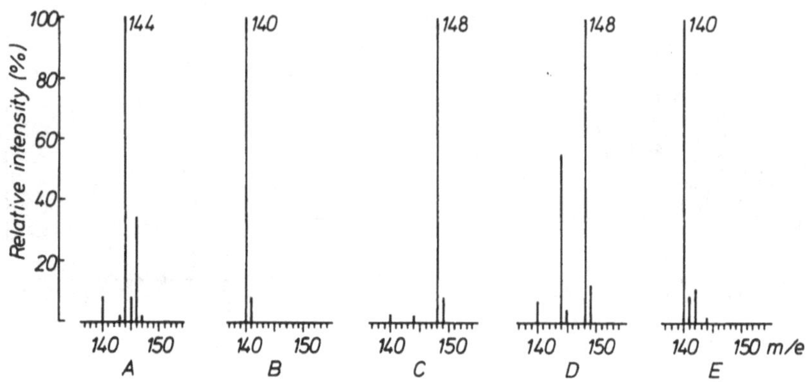

Fig. 4.
Fragments (m/e 140—150 regions containing the parent peaks) of the mass spectra obtained for the compounds accumulating in 0.1 M phosphate buffer, pH 6.9, during 3-chloroaniline turnover by aniline-grown cells of *Rhodococcus* sp. An 117 under the following conditions of oxygen supply: A — control (air atmosphere), B — $^{16}O_2$-atmosphere, C — $^{18}O_2$-atmosphere, D — $^{16}O_2$-$^{18}O_2$-atmosphere, E — air atmosphere/$H_2{}^{18}O$ (in all cases the initial concentrations were: biomass, 1,6 mg of dry weight per milliliter; 3-chloroaniline, 1 mM; glucose, 500 mg/ml)

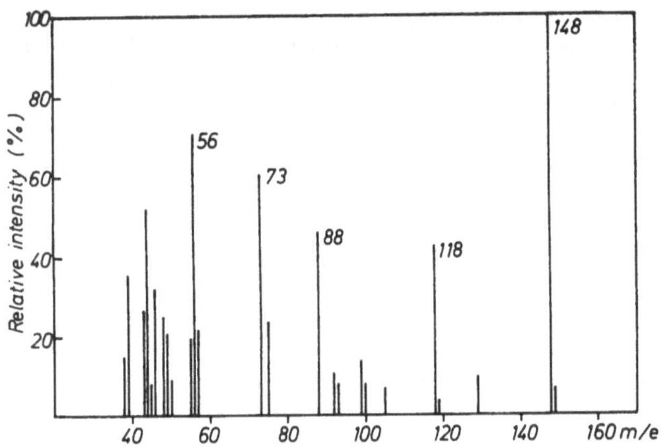

Fig. 5.
Mass spectrum of the compound accumulating during cometabolic 3-chloroaniline turnover by resting aniline-grown An 117 cells in $^{18}O_2$-atmosphere

with $H_2{}^{18}O$ (Fig. 4/E). The mass spectrum of the compound accumulating in $^{18}O_2$-atmosphere, however, showed a parent peak at m/e 148 which clearly indicates the incorporation of four atoms of oxygen-18 (Fig. 4/C and 5).

Of particular interest is variant D (Fig. 4) where accidentally the orginal $^{18}O_2$-atmosphere was gradually changed by air. In this case, in addition to the parent peak at m/e 148, there were smaller peaks at m/e 140 and 144 indicating the incorporation of two atoms each of oxygen-16 and oxygen-18 into the accumulating γ-lactone. The absence of peaks at m/e 142 and 146 suggests the incorporation of two whole molecules of oxygen rather than of four separate oxygen atoms and is consistent with two dioxygenase-mediated reactions rather than with the successive action of monooxygenases. This is in agreement with our finding that aniline-grown An 117 cells converted phenol (1 mM) and 2-aminophenol (1 mM) in the presence of chloramphenicol (150 µg/ml) with rates not exceeding 3% of that seen with aniline (1.5 mM).

Discussion

Cometabolic degradation of 3-chloroaniline in air atmosphere by resting aniline-grown cells of *Rhodococcus* sp. An 117 was found to be characterized by a transient accumulation of 4-chlorocatechol (SCHUKAT *et al.* 1983). In oxygen atmosphere, however, no appreciable amounts of this metabolite were observed, probably due to an activation of the ring-cleavage reaction in An 117 cells. Instead of 4-chlorocatechol, a non-chlorinated compound accumulated which was identified as the γ-lactone of β-hydroxymuconic acid (γ-carboxymethylene-$\Delta^{\alpha\beta}$-butenolide). This metabolite was also observed when the cometabolic process occurred in air atmosphere (AL-MOFARJI *et al.* in preparation) suggesting that elimination of the organically bound chlorine substituent during cometabolic degradation of 3-chloroaniline by *Rhodococcus* sp. An 177 is associated with lactonization of the 3-chloromuconic acid, a dechlorination mechanism found in pseudomonads (EVANS *et al.* 1971, GAUNT and EVANS 1971, KNACKMUSS *et al.* 1976).

The accumulating γ-lactone of β-hydroxymuconic acid contained four atoms of atmospheric oxygen as shown by studies with $^{18}O_2$ and $H_2{}^{18}O$. The same amount of atmospheric oxygen was incorporated into the *cis, cis*-muconic acid during antharnilate conversion by enzyme preparations from *Pseudomonas fluorescens* ATCC 11250) containing anthranilate 1,2-dioxygenase (deaminating, decarboxylating; EC 1.14. 12.1) as well as certain amounts of catechol 1,2-dioxygenase (KOBAYASHI *et al.* 1964). The former enzyme converts anthranilate into catechol by mediating the formation of a cyclic peroxide intermediate, its NADH-consuming reductive cleavage, and concomitant release of one mole each of ammonia and CO_2 per mole of the substrate utilized (TANIUCHI *et al.* 1964, KOBAYASHI and HAYAISHI 1970).

When the cometabolic degradation of 3-chloroaniline by *Rhodococcus* sp. An 117 was performed in the presence of both $^{18}O_2$ and $^{16}O_2$ there was no evidence for mixed-isotope incorporation into the γ-lactone indicating that, in addition to the catechol 1,2-dioxygenase present in aniline-grown An 117 cells (KAMINSKI *et al.* 1983), a second dioxygenase is involved in formation of the accumulating metabolite. The latter enzyme is most likely identical with that responsible for the initial conversion of aniline as well as of 2- and 3-chloroaniline by resting aniline-grown An 117 cells (SCHUKAT *et al.* 1983) and corresponds to the dioxygenase catalyzing the NAD(P)H-consuming conversion of aniline into catechol in a *Nocardia* mutant strain (BACHOFER *et al.* 1975). It is different, however, from anthranilate 1,2-dioxygenase (EC 1.14. 12.1) as seen from the failure of aniline-grown An 117 cells to oxidize anthranilate.

Acknowledgements

We are grateful to V. G. SAKHAROVSKY (Institute of Biochemistry and Physiology of Microorganisms, Pushchino) and his collegues for recording the ^{1}H-NMR spectra.

References

BACHOFER, R., LINGENS, F. and SCHÄFER, W., 1975. Conversion of aniline into pyrocatechol by a *Nocardia* sp.: incorporation of oxygen-18. FEBS Letters, **50**, 288.

EVANS, W. C., SMITH, B. S. W., MOSS, P. and FERNLEY, H. N., 1971. Bacterial metabolism of 4 chlorophenoxyacetate. Biochem. J., **122**, 509.

GAUNT, J. K. and EVANS, W. C., 1971. Metabolism of 4-chloro-2-methylphenoxyacetate by a soil pseudomonad. Ring-fission, lactonizing and delactonizing enzymes. Biochem. J., **122**, 543.

KAMINSKI, U., JANKE, D., PRAUSER, H. and FRITSCHE, W., 1983. Degradation of aniline and monochloroanilines by *Rhodococcus* sp. An 117 and a pseudomonad: a comparative study. Z. allg. Mikrobiol., **23**, 235.

KNACKMUSS, H.-J., HELLWIG, M., LACKNER, H. and OTTING, W., 1976. Cometabolism of 3-methylbenzoate and methylcatechols by a 3-chlorobenzoate utilizing *Pseudomonas*: accumulation of (+)-2,5-dihydro-4-methyl- and (+)-2,5-dihydro-2-methyl-5-oxo-furan-2-acetic acid. Eur. J. Appl. Microbiol., **2**, 267.

KOBAYASHI, S. and HAYAISHI, O., 1970. Anthranilic acid conversion to catechol (*Pseudomonas*). In: Methods in Enzymology, **17a**, 505.

KOBAYASHI, S., KUNO, S., ITADA, N., HAYAISHI, O., KOZUKA, S. and OAE, S., 1964. O^{18} studies on anthranilate hydroxylase — a novel mechanism of double hydroxylation. Biochem. biophys. Res. Commun., **16**, 556.

REBER, H., HELM, V. and KARANTH, N. G. K., 1979. Comparative studies on the metabolism of aniline and chloroanilines by *Pseudomonas multivorans* strain An 1. Eur. J. Appl. Microbiol. Biotechnol., **7**, 181.

SCHUKAT, B., JANKE, D., KREBS, D. and FRITSCHE, W., 1983. Cometabolic degradation of 2- and 3-chloroaniline because of glucose metabolism by *Rhodococcus* sp. An 117. Current Microbiol., **9**, 81

SKRYABIN, G. K., GANBAROV, KH. G. and GOLOVLEVA, L. A., 1976. Peripheral metabolism of isomeric xylenes by a culture of *Pseudomonas aeruginosa*. Mikrobiologiya, **45**, 951.

SUROVTSEVA, E. G., VOLNOVA, A. I. and SHATSKAYA, T. YA., 1980. Degradation of monochloro-substituted anilines by *Alcaligenes faecalis*. Mikrobiologiya, **49**, 351.
TANIUCHI, H., HATANAKA, M., KUNO, S., HAYAISHI, O., NAKAJIMA, M. and KURIHARA, N., 1964. Enzymatic formation of catechol from anthranilic acid. J. biol. Chemistry, **239**, 2204.
YOU, I.-S. and BARTHA, R., 1982. Metabolism of 3,4-dichloroaniline by *Pseudomonas putida*. J. agric. Food Chem., **30**, 274.
ZEYER, J. and KEARNEY, P. C., 1982. Microbial degradation of para-chloroaniline as sole carbon and nitrogen source. Pesticide Biochem. Physiol., **17**, 215.

Mailing address: Dr. D. JANKE, Zentralinstitut für Mikrobiologie, und experimentelle Therapie der AdW, DDR 6900 Jena, Beutenbergstr. 11

Buchbesprechungen

E. C. Friedberg and Ph. C. Hanawalt (Editors), DNA Repair — a Laboratory Manual of Research Procedures, Volume 2. XV + 313 S., 90 Abb., 15 Tab. New York-Basel 1983. Marcel Dekker, Inc. SFr. 132.00. ISBN 0-8247-1805-4.

Die Besprechung der Teile A und B des 1. Bandes erfolgte in der Z. allg. Mikrobiol. 22, 279 und 602 (1982). Die dort gegebenen allgemeinen Bemerkungen sind auch für den 2. Band voll gültig. Die 20 Beiträge sind in 5 Abschnitten zusammengefaßt:

1. Zur enzymatischen Bestimmung von DNA-Schäden werden folgende Enzyme behandelt; S 1 Endonuclease, BAL 31 Nuclease, O^6, Methylguanin-DNA Methyltransferase, 3-Methyladenin-DNA Glykosylase (von *Escherichia coli* und aus humaner Placenta).
2. Bei der Quantifizierung von Pyrimidin-Dimeren in DNA wird die Induktion von Pyrimidin-Dimeren, die Messung von Thymin-enthaltenden Pyrimidindimeren durch HPLC, sowie die Messung von Pyrimidin-Dimeren in Uracil-markierter und nichtmarkierter DNA vorgestellt.
3. Für die Erfassung von Strukturveränderungen in DNA werden die Methoden Viscoelastometrie, Markierung mit Fluoreszenzfarbstoffen, ^{32}P Markierung nach Behandlung mit methylierenden Substanzen, sowie Interstrang crosslink Bildung durch Mitomycin C beschrieben.
4. Der Einsatz von definierten DNA-Sequenzen als Indikatoren für DNA Veränderungen ist dargestellt für DNA Damage, Mutagenese und für das Studium der Genauigkeit der DNA Polymerase.
5. Die Klonierung und Expression von Reparaturgenen ist Gegenstand der abschließenden Beiträge. Behandelt sind von *Escherichia coli* die folgenden Gene: *phr, uvr A, uvr B, uvr C* und *rec A*.

Der vorliegende Band ergänzt den 1. Band und wird wegen der klaren und umfassenden Darstellung bei den Nutzern wieder eine gute Aufnahme finden. H.-E. Jacob (Jena)

E. S. Golub, Die Immunantwort — Einführung in die Immunologie. XII + 306 S., 120 Abb., 32 Tab. Berlin - Heidelberg - New York 1982. Springer-Verlag. DM 29,80.

Dieses Buch ist eine Übersetzung der 1981 erschienenen 2. Ausgabe des amerikanischen Orginals „The Cellular Basis of the Immune Response", und die Übersetzer hätten gut daran getan, den Orginaltitel unverändert beizubehalten. Tatsächlich handelt es sich nicht um eine Einführung in die Immunologie im üblichen Sinne, sondern anhand der Beschreibung fundamentaler Experimente wird der Leser hingeführt, wie es zur Auffindung der modernen Erkenntnisse über die molekularbiologischen und vor allem die zellulären Vorgänge bei der Immunantwort kam. Darin liegt ganz eindeutig der Schwerpunkt der Darstellung und der Vorzug dieses Buches, während andere Aspekte der Immunbiologie, z. B. auch der biochemische Hintergrund der Prozesse nicht mit der gleichen Ausführlichkeit und Sorgfalt dargestellt werden. Hier vermißt man dann auch eine Reihe moderner Erkenntnisse (z. B. Rolle der Lymphokine, Interleukine *etc.*), und es passieren Fehler: z. B. führt die Bindung der Antikörper an Mastzellen allein noch nicht zur Degranulierung (p. 288); die Abläufe bei der Komplementaktivierung werden ungenau dargestellt, nicht alle C-Komponenten reagieren mit Antikörpern und auch nur mit IgG und IgM (p. 286), die Existenz eines alternativen Komplementweges wird nicht erwähnt. Einige andere Fehler dürften den Übersetzern anzulasten sein: z. B. „Kohlenwasserstoffe" der Antikörper (p. 202 und 206) und „zyanogene Bromidpeptide" (p. 245 und 246). Insgesamt kann das Buch wohl als eine hilfreiche Unterstützung für das Verständnis der zellulären Aspekte der Immunantwort empfohlen werden, als alleinige Grundlage für eine Einführung in die Immunbiologie ist es aber nicht ausreichend.

M. Wagner (Jena)

Zeitschrift für allgemeine Mikrobiologie 24 (1984) 4, 261–267

(Delta Institute for Hydrobiological Research, 4401 EA Yerseke, The Netherlands)

The uptake capacity for organic compounds of two heterotrophic bacterial strains at carbon-limited growth[1])

A. B. J. SEPERS

(Eingegangen am 21. 9. 1983)

Two heterotrophic bacterial strains (HIS 42 and HIS 53) were grown in a chemostat at different specific growth rates with histidine (HIS 42) or aspartate (HIS 53) as the growth-limiting nutrient. In these cultures the concentration of the enzyme systems necessary for the transport of organic compounds and their subsequent utilization in respiratory and biosynthetic pathways was measured. The uptake activity increased with an increase of the specific growth rate, as far as the growth-limiting nutrient was concerned. Considering the other applied substrates, an increase of the specific growth rate effected increasing uptake activities (HIS 53: asparagine, glutamate and glutamine) as well as decreasing uptake activities (HIS 53: glucose, propionate, butyrate, lactate, glycollate, glycerate, citrate, α-ketoglutarate, pyruvate and glycerol). However, for the majority of the applied substrates no distinct relationship between the specific growth rate and the uptake activity could be assessed.

Bacteria regulate their enzyme levels in relation with the environmental conditions. One of the controlling parameters is the specific growth rate of the organism. DEAN (1972) presented in a review paper examples of different relationships between the specific growth rate and the concentration of several enzyme systems. So increasing growth rates were reported to bring about increasing as well as decreasing enzyme concentrations dependent on the enzyme system considered. Also enzyme levels passing through a maximum or minimum level at a given growth rate were described and in some other cases a constant enzyme level irrespective of the growth rate was shown.

MATIN *et al.* (1976) reported an increase in the concentration of intermediary enzymes with decreasing growth rates. A *Pseudomonas* and a *Spirillum* species were grown in a chemostat culture at a lactate limitation. A decrease of the growth rate was accompanied by an increase in the concentration of lactate dehydrogenase, aconitase and glucose-6P-dehydrogenase. The authors point to the ecological advantage of this type of response. The higher enzyme activities at the lower growth rates enhance the ability of the organism to utilize nutrients which are only present at very low levels.

In this paper experiments are reported dealing with the concentration of enzyme systems necessary for the transport of organic compounds and their subsequent utilization in respiratory and biosynthetic pathways. To this end two heterotrophic bacterial strains (HIS 42 and HIS 53) were grown in a chemostat at different growth rates with histidine (HIS 42) or aspartate (HIS 53) as the growth-limiting nitrient. On samples from these chemostats the enzyme activities were measured.

Materials and methods

The experiments were done with bacterial strains HIS 42 and HIS 53, which were isolated from the freshwater Haringvliet basin in The Netherlands by enrichment in a chemostat culture at a

[1]) Communication no. 286 of the Delta Institute for Hydrobiological Research, 4401 EA Yerseke, The Netherlands

dilution rate of 0.05 h^{-1} and with histidine as the growth-limiting nutrient (SEPERS 1981). The maximum specific growth rate (μ_{max}) of HIS 42 for growth on histidine is 0.20 h^{-1}; the μ_{max} of HIS 53 for growth on aspartate is 0.36 h^{-1}.

HIS 42 was grown in a histidine-limited chemostat at specific growth rates of 0.01, 0.05 and 0.1 h^{-1}. Samples taken from the chemostat were used to determine the uptake kinetics for histidine, glycine, leucine, aspartate, an algal protein hydrolysate, glucose and acetate. The methodology for the determination of these uptake kinetics was principally that of HOBBIE and CRAWFORD (1969) with some minor modifications as described by SEPERS and VAN ES (1979). The maximum uptake rate, as measured with this technique, is proportional to the concentration of the enzyme systems necessary for the uptake of the considered substrate and the subsequent utilization in respiratory and biosynthetic pathways.

The uniformly ^{14}C-labelled organic compounds were purchased from the Radiochemical Centre, Amersham. The stock solution was diluted in 0.001 N HCl to an activity of 1 μCi/ml. The diluted solution was sterilized by heat, and thereafter the pH was adjusted to 7.5 ± 0.2. The substrate concentration was determined by measuring the radioactivity of the same sample as applied in the experiments. From the radioactivity of this sample and the specific activity as stated by the factory, the substrate concentration could be calculated. In Table 1 the range of the applied substrate concentrations is presented.

Table 1

Survey of the substrate concentrations as applied in the uptake experiments with HIS 42

Substrate	Product code Radiochemical Centre, Amersham	Substrate concentration
Glycine	CFB 66	16—266 nM
Leucine	CFB 67	5— 88 nM
Histidine	CFB 140	5— 90 nM
Aspartate	CFB 64	8—137 nM
Algal protein hydrolysate	CFB 25	33—535 ng-at C/l
Glucose	CFB 96	6—103 nM
Acetate	CFA 229	36—559 nM

HIS 53 was grown in a chemostat with aspartate as the growth-limiting substrate at the specific growth rates of 0.001, 0.003 and 0.01 h^{-1}, and in batch culture at the specific growth rate of 0.34 h^{-1}. Samples taken from these cultures were used to determine the uptake kinetics of 39 organic compounds by measurement of the oxygen consumption rate with a Biological Oxygen Monitor (Yellow Spring Instruments, Model 53 SA). The oxygen consumption rate measured at saturating substrate concentrations is proportional to the concentration of the enzyme systems necessary for the uptake and respiration of the considered substrate.

The oxygen uptake was measured after the addition of 1 mM of substrate. Some preliminary experiments demonstrated that at this substrate concentration the maximum uptake rate was measured. The measured oxygen consumption rate was converted to an uptake rate for the substrate in question assuming a full respiration of the substrate.

For the samples from the chemostat culture of HIS 53 with a dilution rate of 0.001 h^{-1} a different procedure was followed. The sample was transferred into 6 oxygen bottles with a volume of about 11 ml. Two of these bottles were fixed immediately, so as to determine the initial oxygen concentration. Another two bottles were incubated for 3–4 hours after the addition of substrate and thereafter fixed. The endogenous respiration was determined by the incubation of two bottles without added substrate. The oxygen content was measured according to the WINKLER method (STRICKLAND and PARSONS 1968). The oxygen consumption rate was calculated from the difference between the initial oxygen content and the oxygen content after incubation. The oxygen consumption values were corrected for the endogenous respiration.

Before the measurement of the oxygen uptake rate in samples from the batch culture of HIS 53 on aspartate, the samples were filtered through a 0.2 μm polycarbonate filter (Uni-Pore membrane filters, BIO-RAD Laboratories) to remove the residual aspartate. The bacteria on the filter were washed twice with medium without aspartate and then resuspended in aspartate free medium. This suspension was used for the determination of the respiratory capacity. In the chemostat cultures of HIS 53 with a dilution rate of 0.001, 0.003 and 0.01 h^{-1} aspartate could not be detected in the culture fluid. Therefore, samples from these cultures could be used without filtration.

The continuous cultures with a dilution rate of 0.01 h^{-1} and less were run in a fermentor as described by HARDER *et al.* (1974); the volume of the cultures grown at dilution rates of 0.001 and 0.003 h^{-1} was 7—8 l and of the cultures grown at a dilution rate of 0.01 h^{-1} 1.5 l. For the other continuous cultures a New Brunswick BioFlo Model C30 fermentor with a culture volume of 350 ml was used. All experiments were carried out on chemostat cultures in steady state.

The batch cultures of HIS 53 were run in 300 ml ERLENMEYER flasks with 100 ml of medium. These cultures were incubated in an incubator shaker.

The composition of the culture medium was (g/l): KH_2PO_4, 0.1; $MgSO_4 \cdot 7\,H_2O$, 0.2; $CaCl_2 \cdot 2\,H_2O$, 0.1; NaCl, 1.0; Na-aspartate (HIS 53), 0.17 (1 mM) or histidine (HIS 42), 0.15 (1 mM); 1 ml of a trace elements solution and 1 ml of an iron citrate solution ($C_6H_5FeO_7 \cdot 5\,H_2O$, 1.5 g/l), were added per liter of medium. The composition of the trace elements solution was (mg/l): H_3BO_3, 564; $CoCl_2 \cdot 6\,H_2O$, 202; $ZnSO_4 \cdot 7\,H_2O$, 441; $CuSO_4 \cdot 5\,H_2O$, 19.5; $MnCl_2 \cdot 4\,H_2O$, 90; $Na_2MoO_4 \cdot 2\,H_2O$, 30. The pH of the medium was 7.5 and the growth temperature was 18 °C.

Protein determinations were done according to BRADFORD (1976) with the BIO-RAD Protein Assay Reagent Concentrate after disintegration of the bacterial cells with a MSE Ultrasonic Disintegrator (Model PG 100; 5 min at an amplitude of 10—12 micron). Aspartate was measured enzymatically, as described by BERGMEYER *et al.* (1975); histidine was determined according to WOIWOD (1949). Organic carbon was measured with the Total Carbon System of the Oceanography International Corp.

Results

The results of the uptake experiments with HIS 42 are shown in Table 2. An increase of the growth rate was accompanied by an increase in the maximum uptake rate, as far as the growth-limiting substrate (histidine) is concerned. The maximum uptake rate of the other substrates was nearly independent of the specific growth rate.

A possible relationship between the specific growth rate and the concentration of an enzyme system may be more pronounced in organisms which are able to grow over a wider range of specific growth rates. Thus for the next series of experiments HIS 53 was chosen growing in a chemostat with aspartate as the growth-limiting nutrient. The maximum specific growth rate of HIS 53 at these culture conditions is 0.36 h^{-1}, whereas the μ_{max} of HIS 42 for growth on histidine is 0.20 h^{-1}.

The results of the experiments measuring the overall activity of the uptake and respiratory systems of HIS 53 on 39 organic compounds, are presented in Table 3. The respiratory activity for aspartate, asparagine, glutamate and glutamine increased with increasing growth rate. It should be noted that at the lowest specific growth rate of 0.001 h^{-1} no glutamine uptake could be detected.

Table 2

The uptake of several organic compounds by HIS 42 grown on histidine at different specific growth rates

Specific growth rate (h^{-1})	Maximum uptake rate of substrate (μmol mg $C^{-1}h^{-1}$)						
	Histidine	Aspartate	Leucine	Glycine	Amino acid mixture	Glucose	Acetate
0.01	0.54	1.59	0.03	0.24	3.84[a]	0.04	0.14
0.05	1.54	2.10	0.03	0.30	5.59	0.04	0.11
0.10	1.52	1.59	0.02	0.18	5.39	0.03	0.27
0.12	2.04						
0.13	3.21						

[a] Maximum uptake rate in μg-at C mg $C^{-1}h^{-1}$

Table 3

The uptake of organic compounds by HIS 53 grown on aspartate at different specific growth rates.
The measured oxygen uptake rates were converted to substrate uptake rates assuming a full respiration of the substrate; — oxygen uptake rate less than 5 µmoles O_2 l^{-1} h^{-1} (at $\mu > 0.003$ h^{-1}) or less than 3 µmoles O_2 l^{-1} h^{-1} (at $\mu = 0.001$ h^{-1}); + oxygen uptake rate between 5 and 20 µmoles O_2 l^{-1} h^{-1} (at $\mu > 0.003$ h^{-1}) or between 3 and 7 µmoles O_2 l^{-1} h^{-1} (at $\mu = 0.001$ h^{-1}). Parentheses indicate number of measurements if the uptake experiments were repeated.

Specific growth rate (h^{-1})	Maximum uptake rate of substrate (µmol mg protein^{-1} h^{-1})							
	Glycine	Alanine	Valine	Isoleucine	Leucine	Lysine	Arginine	Histidine
0.001	0.39	0.23	+	+	+	—	—	— (2)
0.003	0.56	0.37	+	0.07	+	—	+	0.14 (2)
0.01	1.12 (2)	0.37 (3)	+	0.09	+	+ (2)	+ (2)	0.18 (4)
0.34	—	+	—	—	—	—	—	—

Specific growth rate (h^{-1})	Maximum uptake rate of substrate (µmol mg protein^{-1} h^{-1})						
	Proline	Serine	Threonine	Aspartate	Asparagine	Glutamate	Glutamine
0.001	0.13	0.24	—	0.38 (7)	0.30	0.36	—
0.003	+	0.60 (2)	0.20	0.46 (11)	0.68	0.36	0.31
0.01	0.12 (2)	0.74 (2)	—	1.53 (16)	1.52 (3)	0.97 (2)	1.01
0.05				1.90 (7)			
0.10				2.64 (6)			
0.12				2.12 (6)			
0.34	—	+	—	4.64 (11)	3.41	2.62	2.77

Specific growth rate (h^{-1})	Maximum uptake rate of substrate (µmol mg protein^{-1} h^{-1})							
	Methionine	Tryptophan	Phenyl-alanine	Ornithine	Glucose	Sucrose	Starch	Acetate
0.001	+	+	+	0.12	0.68 (2)	+	—	0.39
0.003	+	—	+	+	+	—	—	+
0.01	+	—	+ (2)	0.11 (2)	+	—	—	1.16 (4)
0.34	—	—	—	—	—	—	—	+

Specific growth rate (h^{-1})	Maximum uptake rate of substrate (μmol mg protein^{-1} h^{-1})							
	Propionate	Butyrate	Oxalate	Malonate	Succinate	Fumarate	Glutarate	Malate
0.001	0.27	0.14	+	—	0.52	—	—	—
0.003	+	—	—	—	—	—	—	—
0.01	—	—	—	—	—	—	—	—
0.34	—	—	—	—	4.21	—	—	—

Specific growth rate (h^{-1})	Maximum uptake rate of substrate (μmol mg protein^{-1} h^{-1})							
	Lactate	Glycollate	Glycerate	Citrate	α-ketoglutarate	Pyruvate	Glycerol	Benzoate
0.001	0.60	0.88	0.25	0.14	0.35	1.07	0.61	+
0.003	+		—	—	—	+	—	—
0.01	—	—	—	—	—	0.70 (6)	—	—
0.34	—	—	—	—	—	+	—	—

For glucose, propionate, butyrate, lactate, glycollate, glycerate, citrate, α-ketoglutarate, pyruvate and glycerol the highest respiratory activity was found at the lowest specific growth rate applied (0.001 h^{-1}). At the higher growth rates the respiratory activity for these substrates was very low or completely absent. For the other substrates no clear-cut relation between the respiratory activity and the specific growth rate could be demonstrated.

Discussion

The experiments showed that the uptake capacity increased with an increase of the specific growth rate as far as the growth-limiting substrate is concerned (HIS 42, histidine; HIS 53, aspartate). This increase can be explained by the adjustment of the physiology of the organisms to the higher growth rates at cultivation in a chemostat culture.

An increase of the specific growth rate can result in increasing uptake activities (HIS 53: asparagine, glutamate and glutamine), as well as in decreasing uptake activities (HIS 53: glucose, propionate, butyrate, lactate, glycollate, glycerate, citrate, α-ketoglutarate, pyruvate and glycerol). However, for the majority of the applied substrates no distinct relationship between the specific growth rate and the uptake activity could be assessed.

Summing up the gathered data, it appeared that the organisms used in the reported experiments can vary the concentration of a number of enzyme systems in connection with the specific growth rate. In a multi-enzyme system the saturation constant of an enzyme is a function of the concentration of all the involved enzymes (KRAMBECK 1979). Consequently the organisms used in this study may be able to regulate their saturation constants by adapting their relative enzyme concentrations. As in the natural aquatic environment the growth of heterotrophic bacteria is normally limited by organic carbon, some such regulation of the affinity for the growth-limiting nutrient will have competitive value for the organism, if the affinity increases at the lower substrate concentrations.

The uptake capacity of HIS 53 was tested on 39 organic compounds. A comparison between the uptake activity measured at the low specific growth rate of 0.001 h^{-1} and the uptake activity at the specific growth rate of 0.34 h^{-1} shows that at the low growth rate 29 organic compounds were taken up and respired (the uptake activity on 9 substrates was very low, indicated by + in Table 3), whereas at the high growth rate only 9 substrates were respired. This can be considered as an indication to a simultaneous uptake of organic compounds by heterotrophic bacteria in their natural environment. Referring to the low growth rates normally occurring in nature, this character contributes to the survival value of these organisms.

Acknowledgements

Thanks are due to MARY CAPPENDIJK, ANJA LUTEYN and F. W. MELISSEN for skilfull technical assistance.

References

BERGMEYER, H. U., BERNT, E., MÖLLERING, H. and PFLEIDERER, G., 1975. L-Aspartate and L-asparagine. In: Methods of Enzymatic Analysis, Vol. IV (edited by H. U. BERGMEYER). Academic Press New York. pp. 1696–1700.

BRADFORD, M. M., 1976. A rapid and sensitive method for the quantitation of microgram quantities of protein utilizing the principle of protein-dye binding. Anal. Biochem., **72**, 248–254.

DEAN, A. C. R., 1972. Influence of environment on the control of enzyme synthesis. J. appl. Chem. Biotechnol., **22**, 245–259.

HARDER, W., VISSER, K. and KUENEN, J. G., 1974. Laboratory fermenter with an improved magnetic drive. Lab. Pract., **23**, 644–645.

HOBBIE, J. E. and CRAWFORD, C. C., 1969. Bacterial uptake of organic substrate: new methods of study and application to eutrophication. Verh. Internat. Verein. Limnol., **17**, 725–730.

KRAMBECK, C., 1979. Applicability and limitations of the MICHAELIS-MENTEN equation in microbial ecology. Arch. Hydrobiol. Beih. Ergebn. Limnol., **12**, 64–76.

MATIN, A., GROOTJANS, A. and HOGENHUIS, H., 1976. Influence of dilution rate on enzymes of intermediary metabolism in 2 freshwater bacteria grown in continuous culture. J. Gen. Microbiol., **94**, 323–332.

SEPERS, A. B. J., 1981. Diversity of ammonifying bacteria. Hydrobiologia, **83**, 343–350.

SEPERS, A. B. J. and VAN ES, F. B., 1979. Heterotrophic uptake experiments with ^{14}C-labeled histidine in a histidine-limited chemostat. Appl. Environ. Microbiol., **37**, 794–799.

STRICKLAND, J. D. H. and PARSONS, T. R., 1968. A practical handbook of seawater analysis. Fish. Res. Bd Can. Ottawa. Bulletin 167. 311 pp.

WOIWOD, A. J., 1949. A method for the estimation of microamounts of amino nitrogen and its application to paper partition chromatography. Biochem. J., **45**, 412–417.

Mailing address: A. B. J. SEPERS, Delta Institute for Hydrobiological Research, Vierstraat 28, 4401 EA Yerseke, The Netherlands

Zeitschrift für allgemeine Mikrobiologie 24 (1984) 4, 268

Buchbesprechungen

R. W. HENDRIX, J. W. ROBERTS, F. W. STAHL and R. A. WEISBERG, Lambda II (Monograph 13). 694 S., 79 Abb. 24 Tab. Cold Spring Harbor 1983. Cold Spring Harbor Laboratory. $ 93.60. ISBN 0-87969-150-6.

Zwölf Jahre nach Herausgabe einer fast 800 S. umfassenden Monographie über den „Bakteriophagen Lambda“ durch A. D. HERSHEY (1971) legen die Autoren mit „Lambda II‘ ein Buch vor, das in 17 Übersichtsartikeln den enormen Wissensfortschritt auf allen Gebieten der Biologie dieses Coliphagen darstellt. Die Forschung am Phagen Lambda hat zu Fragen der Genregulation und Genfunktion fundamentale Beiträge geliefert und liefert sie noch. Das machen die einzelnen Beiträge aus der Feder der Autoren deutlich, die zu den erfolgreichsten Forschern auf diesem Gebiet zählen. U. a. werden Beiträge zu folgenden Problemen der Lambda-Biologie gebracht: Lytische Phagenentwicklung, Repressorsynthese u. deren Kontrolle, Kontrolle von Integration und Exzision des Prophagen, lysogene Induktion, DNA-Replikation, generelle u. ortsspezifische Rekombination, Assemblierung der Kopfhülle, DNA-Verpackung, Schwanzsynthese u. Injektion, Evolution des Phagen.

Dem Anfänger auf dem Gebiet der Lambda-Forschung werden die Beiträge von W. ARBERS „Einführung in die Lambda-Biologie“ u. „Experimentelle Methoden“ von besonderem Nutzen sein, dem an Gentechnik Interessierten der Artikel von N. MURRAY über die Verwendung von Lambda-Vektoren zur Gen-Klonierung. Dem letzteren Anliegen dienen auch die im Anhang enthaltenen Daten: die komplette Basensequenz des Phagengenoms (48. 514 bp), eine darauf basierende molekulare Karte mit allen Gen- u. Restriktase-Spaltorten sowie eine Liste der Lambda-Vektoren mit den jeweils auf die Basensequenz bezogenen Restriktase-Spaltorten.

Das Buch kann ohne Einschränkung dem Studenten der Molekularbiologie ebenso wie dem versierten „Lambdologen“ bestens empfohlen werden. S. KLAUS (Jena)

H. KINDL and P. B. LAZAROW (Editors), Peroxisomes and Glyoxysomes (Annals of the New York Academy of Sciences, Volume 386). 550 S., 292 Abb., 64 Tab. New York 1982. The New York Academy of Sciences. ISBN 0-89766-162-1.

Der vorliegende Tagungsband gibt eine gute Übersicht des gegenwärtigen Kenntnisstandes zur Ultrastruktur, Biogenese und Funktion der Peroxisomen und Glyoxysomen in pflanzlichen bzw. tierischen Zellen sowie der „Microbodies“ in eukaryotischen Mikroorganismen. Eine große Anzahl der Beiträge beschäftigt sich mit der Charakterisierung von Enzymen dieser Zellorganelle, deren Funktion, sowie mit der Verknüpfung von Stoffwechselwegen, die in diesen Organellen ablaufen, mit dem Metabolismus der Zelle bzw. anderer Zellorganelle. Dabei wird deutlich, daß die Ebene des bloßen Beschreibens solcher Strukturen bereits längst verlassen wurde und gegenwärtig immer mehr die Aufklärung deren komplexen Funktion in den Vordergrund tritt.

Die Untersuchung komplexer Stoffwechselprozesse und die Aufklärung deren Regulation in der eukaryotischen Zelle erscheint ohne Beachtung dieser Zellstrukturen wenig erfolgreich. Deshalb wird dieses Buch nicht nur den unmittelbar auf diesem Gebiet tätigen Wissenschaftlern empfohlen, sondern auch all denen, die Stoffwechselwege oder die Rolle und Funktion einzelner Enzyme in eukaryotischen Zellen untersuchen. G. BARTH (Jena)

Zeitschrift für allgemeine Mikrobiologie 24 (1984) 4, 269–279

(Akademie der Wissenschaften der DDR, Forschungszentrum für Molekularbiologie und Medizin, Institut für Biochemie der Pflanzen, Halle/Saale)

Wirkung von Metalaxyl auf die Synthese von RNA, DNA und Protein in *Phytophthora nicotianae*

R. Wollgiehn, E. Bräutigam, B. Schumann und D. Erge

(Eingegangen am 9. 9. 1983)

Metalaxyl is used to control diseases caused by fungi of the order of the *Perenosporales*. We investigated the action of this fungicid eon nucleic acid and protein synthesis in liquid cultures of *Phytophthora nicotianae*.

The uptake of ^{32}P, ^{3}H-uridine, ^{3}H-thymidine and ^{14}C-leucine as precursors of nucleic acid and protein synthesis by the mycelium was not inhibited by metalaxyl.

RNA synthesis as indicated by ^{3}H-uridine incorporation was strongly inhibited (about 80%) by 0.5 μg/ml of metalaxyl. The inhibition was visible already few minutes after addition of the toxicant. Since the inhibition of incorporation of ^{3}H-thymidine into DNA and of ^{14}C-leucine into protein became significant 2–3 hours later, we conclude that metalaxyl primarily interfers with RNA synthesis.

Synthesis of ribosomal RNA is more affected (more than 90%) than that of tRNA (about 55%) and poly(A)-containing RNA. Since in the presence of actinomycin, in contrast to metalaxyl, protein synthesis is inhibited immediately as a consequence of complete inhibition of RNA synthesis and of the short life-time of mRNA, it is also evident that mRNA synthesis is less strongly inhibited, at least during the early period of metalaxyl action.

The molecular mechanism of metalaxyl inhibition of the transcription process remains open. The fungicide did not inhibit the activity of a partially purified RNA polymerase isolated from the fungus. On the other hand, the RNA synthesis (^{14}C-UTP-incorporation) by a cell homogenate and by isolated nuclear fractions was inhibited significantly. Possibilities of the molecular action of metalaxyl are discussed.

The RNA synthesis of some plant systems (cell cultures of *Lycopersicon peruvianum*, isolated nuclei from the same cell cultures, purified RNA polymerase from *Spinacia oleracea* chloroplasts) was not inhibited by metalaxyl, not even at high concentrations.

Bereits seit mehreren Jahren werden Acylalanin-Fungicide (Metalaxyl, Furalaxyl) erfolgreich zur Bekämpfung von pathogenen Pilzen aus der Ordnung der *Peronosporales* (*Phytophthora*, *Pythium*, *Plasmopara*) eingesetzt (Schwinn *et al.* 1977, Urech *et al.* 1977). Obwohl mehrere detaillierte Untersuchungen über die Spezifität und über physiologische Aspekte der Metalaxyl-Wirkung vorliegen (Staub *et al.* 1978, 1980 Kerkenaar u. Sijpesteijn 1981), ist über den molekularen Wirkungsmechanismus dieses Fungicids noch wenig bekannt. Die bisherigen Befunde ergaben, daß Atmung und Zellwandsynthese nicht primär beeinflußt werden (Kerkenaar u. Sijpesteijn 1981, Benson 1979, Fisher u. Hayes 1982), wogegen Einbau-Versuche mit radioaktiven Precursoren darauf hindeuten, daß der Wirkungsmechanismus in der Beeinflussung von Replikation oder Genexpression zu suchen ist. Während Kerkenaar (1981) bei *Pythium splendans* eine Hemmung der RNA- und DNA-Synthese, nicht aber der Protein-Synthese beobachtete, fanden Fisher und Hayes (1982) bei *Pythium* und *Phytophthora* sowohl eine Hemmung der Nucleinsäure- als auch der Protein-Synthese durch Metalaxyl. Beide Untersuchungen ließen auf einen primären Angriffspunkt im RNA-Stoffwechsel schließen. Bestätigt und ergänzt wird diese Annahme durch neuere Befunde von Davidse (1983) an *Phytophthora*, wonach Metalaxyl *in vivo* und in isolierten Zellkernen bevorzugt die Synthese Poly(A)-freier RNA hemmt, die Aktivität isolierter RNA-Polymerasen jedoch nicht beeinflußt.

Die vorliegenden Untersuchungen befassen sich mit der Wirkung von Metalaxyl auf die RNA-, DNA- und Protein-Synthese in *Phytophthora nicotianae*. Die Ergebnisse zeigen, daß das Fungicid primär in die RNA-Synthese eingreift.

Material und Methoden

Die Versuche wurden an *Phytophthora nicotianae* durchgeführt. Verwendete Chemikalien und radioaktiv markierte Verbindungen: Metalaxyl (DL-Methyl-N-(2,6-Dimethylphenyl)-N-(2-Methoxyacetyl)-Alaninat, CGA 48988) von CIBA-GEIGY, Basel; Actinomycin von REANAL, Budapest; Ampicillin von PHARMACHIM, Sofia; Acrylamid, N,N'-Methylenbisacrylamid, N,N,N',N'-Tetramethyläthylendiamin, Lauroylsarkosin-Na von SERVA, Heidelberg; Naphthalin-1,5-Disulfonsäure-Di-Na, Formamid, Ammoniumpersulfat von MERCK, Darmstadt; Poly(U)-Sepharose 4B von PHARMACIA, Uppsala; Polyäthylenimin von BDH CHEMICALS, Pool/England; DEAE-Cellulose DE 52 von WHATMAN, Maidstone/England; $Na_2H^{32}PO_4$, trägerfrei, vom Zentralinstitut für Kernforschung, Dresden-Rossendorf; 5-^{3}H-Uridin (814 GBq/mM), 6-^{3}H-Thymidin (1050 GBq/mM), U-^{14}C-L-Leucin (8880 MBq/mM) und U-^{14}C-Uridin-5'-Triphosphat (14,8 GBq/mM) von UVVVR, Prag.

Nährmedien:

1. SAKAI-Agar: 34 g Bohnenmehl, 17 g Hafermehl, 8,5 g Saccharose, 15 g Agar, dest. Wasser ad 1000 ml.
2. NL 962: 50 g Saccharose, 15 g Sojamehl, 10 ml Maisquellwasser, 500 mg KH_2PO_4, 300 mg $MgSO_4 \cdot 7\,H_2O$, 10 mg $FeSO_4 \cdot 7\,H_2O$, 8,8 mg $ZnSO_4 \cdot 7\,H_2O$, dest. Wasser ad 1000 ml, pH 5,4—5,6. Nach 15 min Kochen wurde das Medium filtriert und anschließend sterilisiert.
3. NL 963 (nach FRIES 1938): 100 g Saccharose, 5 g Bernsteinsäure, 1 g NH_4NO_3, 100 mg KH_2PO_4, 300 mg $MgSO_4 \cdot 7\,H_2O$, 100 mg NaCl, 100 mg $CaCl_2$, 10 mg $FeSO_4 \cdot 7\,H_2O$, 5 g Hefeextrakt (DIFCO), dest. Wasser ad 1000 ml, pH 5,4.
4. NL 965: 200 g Saccharose, 15 g Ammoniumcitrat, 1 g $Ca(NO_3)_2$, 300 mg $MgSO_4 \cdot 7\,H_2O$, 10 mg $FeSO_4 \cdot 7\,H_2O$, 30,8 mg $ZnSO_4 \cdot 7\,H_2O$, 100 mg KCl, dest. Wasser ad 1000 ml, pH 5,2—5,4.

Die Lösungen wurden zur Sterilisation 20 min bei 120 °C, NL 965 30 min bei 110 °C im Autoklaven erhitzt.

Kulturtechnik: Die Erhaltung des Pilzstammes erfolgte auf SAKAI-Schrägagarröhrchen. Das Mycel wurde in Abständen von 14—21 Tagen übertragen. Die Kultivation erfolgte bei 24° im Dunkeln.

Zur Submerskultivation wurde die Hälfte des Mycels eines 2—3 Wochen alten Schrägagarröhrchens in 10 ml steriler physiologischer Kochsalzlösung mittels eines Messer-Homogenisators (WARING-Blendor) 0,5 min homogenisiert und anschließend in einen 500-ml-Langhalskolben, der mit 75 ml NL 962 und 50 mg Ampicillin beschickt war, übertragen. Diese Vorkultur wurde 2—3 Tage im Dunkeln bei 24° geschüttelt (FANAL-Rundschüttler, 240 rpm). Die gut gewachsene Vorkultur wurde homogenisiert, und zur Beimpfung der Hauptkultur wurden 7,5 ml Suspension eingesetzt. Als Hauptkultur verwendet man 75 ml NL 963 mit einem Zusatz von 25 mg Ampicillin in einem 500-ml-Langhalskolben. Die Kulturdauer betrug 2—3 Tage. Es ist auch möglich, die Vorkultur mehrfach in NL 962 zu passagieren und dieses Mycel zur Beimpfung der Hauptkultur zu benutzen. Der Zusatz von Ampicillin erwies sich am günstigsten zur Vermeidung von bakteriellen Verunreinigungen, die ohne Zusatz eines Antibioticums in den Kulturen auftraten. Vor jedem Impfschritt wurde die Reinheit der Kultur überprüft. Zur Durchführung der ^{32}P-Markierungsversuche war es erforderlich, die Kulturen vor Versuchsbeginn an Phosphat zu verarmen. Dazu wurde das Mycel der Hauptkultur mittels einer Glasfilternutsche abgesaugt, mit 0,9% NaCl-Lösung gewaschen und nach der Übertragung in 75 ml phosphatfreie NL 965 16 Std. geschüttelt. Um die Resultate vergleichen zu können, wurde dieses Verfahren für alle Versuche angewandt, bei denen radioaktive Tracer eingesetzt wurden.

Radioaktive Markierungen: Zur Gewinnung von ^{32}P-markierten Nucleinsäuren wurden 75 ml Phosphat-verarmter Kultur in NL 965 in 500 ml Kolben 15 min mit Metalaxyl in den angegebenen Konzentrationen vorinkubiert und dann nach Zusatz von 1,5 mg KH_2PO_4 und 80 MBq $Na_2H^{32}PO_4$ weitere 3 Std. bei 24 °C geschüttelt. ^{3}H-Uridin-markierte Nucleinsäuren wurden aus dem Mycel einer 20 ml Kultur isoliert. Nach 15 min Vorinkubation mit Metalaxyl wurde die Kultur nach Zusatz von 0,4 mg KH_2PO_4 und 8 MBq ^{3}H-Uridin 3 Std. geschüttelt. Das Mycel wurde auf Papierfilter abgesaugt, gewaschen und zur Extraktion von Nucleinsäuren verwendet.

Die Einbau-Kinetiken von ^{32}P, ^{3}H-Uridin, ^{3}H-Thymidin und ^{14}C-Leucin wurden wie folgt bestimmt. 15 ml Zellkultur wurden 15 min mit Metalaxyl vorinkubiert. Nach Zusatz von 0,4 mg KH_2PO_4 und 4 MBq $Na_2H^{32}PO_4$ oder 4 MBq ^{3}H-Uridin oder 4 MBq ^{3}H-Thymidin oder 2 MBq

^{14}C-Leucin und 50 µg L-Leucin wurden die Kolben bei 24 °C geschüttelt. Zu den angegebenen Zeiten wurde jeweils 1 ml der Suspension entnommen und in 5 ml kalter 5% Trichloressigsäure pipettiert. Nach einigen Std. wurde das ausgefällte Material auf Filterpapierscheiben abgesaugt, mehrfach mit TES und einmal mit Äthanol gewaschen. Nach dem Trocknen wurde die Radioaktivität im PACKARD Scintillationsspektrometer gemessen.

Nucleinsäure-Extraktion: Das Mycel (*ca.* 1 g Feuchtgewicht aus einer 75 ml Kultur) wurde in flüssigem Stickstoff zerrieben und in 25 ml 100 mM Tris-HCl (pH 9), 150 mM NaCl, 5 mM $MgCl_2$ 6% *p*-Aminosalicylsäure, 1% Naphthalin-1,5-Disulfonsäure aufgenommen. Das mehrfache Deproteinisieren erfolgte jeweils durch Zusatz von 1 Vol. Phenol/Chloroform (1:1), Rühren und Abzentrifugieren. Die Nucleinsäuren wurden mit 3 Vol. Äthanol über Nacht bei −20 °C gefällt, danach abzentrifugiert und nach Lösen in 150 mM NaCl, 15 mM Na-Citrat (1 SSC) erneut mit Alkohol gefällt.

Nucleinsäure-Elektrophorese: Die Polyacrylamidgel-Elektrophorese (2,4% oder 7,5% Polyacrylamid), das Scannen der optischen Dichte bei 260 nm und die Bestimmung der Radioaktivität nach Schneiden der Gele mit ^{3}H-markierten Nucleinsäuren in 1 mm Scheiben oder durch Scannen der Autoradiogramme getrockneter Gele mit ^{32}P-markierten Nucleinsäuren bei 600 nm erfolgte wie früher beschrieben (MUNSCHE u. WOLLGIEHN 1973, WOLLGIEHN *et al.* 1974, SCHIEMANN *et al.* 1978). In den Abbildungen ist die Radioaktivität der ^{32}P-Gele als Relativwert zur Schwärzung der Autoradiogramme angegeben ($O.D._{600\,nm}$).

Isolierung von Poly(A)$^{+}$-RNA: Vor der Trennung in Poly(A)$^{+}$- und Poly(A)$^{-}$-RNA wurden die Nucleinsäuren in Wasser gelöst, mit 9 M LiCl zu einer Endkonzentration von 3 M LiCl versetzt und mehrere Stunden in Eis aufbewahrt. Dabei fallen die hochmolekularen RNA (rRNA, mRNA) aus, während DNA und niedermolekulare RNA in Lösung bleiben. Nach dem Abzentrifugieren wurde die hochmolekulare RNA 3mal mit 80% Äthanol + 0,5 M NaCl gewaschen. Die Isolierung der Poly(A)$^{+}$-RNA erfolgte durch (Poly(U)-Sepharose-Chromatographie wie früher beschrieben (WOLLGIEHN *et al.* 1978). Die RNA wurde zweimal über die Säule gegeben.

Isolierung von Zellkernen aus *Phytophthora*: 20 g Pilzmycel wurden mit flüssigem Stickstoff zerrieben und mit 10 Vol. Puffer A (50 mM Tris-HCl (pH 7,8), 4 mM EDTA, 0,5 g/l Cystein, 1 g/l Rinderserumalbumin, 0,3 M Saccharose) aufgenommen. Das Homogenat wurde 30 min gerührt und durch Mühlensiebgewebe (32 µm Maschenweite) filtriert. Das Filtrat wurde 10 min bei 50 × g zentrifugiert, der Überstand 10 min bei 1300 × g. Das Sediment wurde in Puffer B (50 mM Tris-HCl (pH 7,8), 4 mM EDTA, 5 mM Mercaptoäthanol, 0,3M Saccharose) aufgenommen und 10 min bei 1300 × g zentrifugiert. Die sedimentierte Zellkernrohfraktion wurde zur weiteren Reinigung in 15 ml Puffer B suspendiert und auf 50 ml eines kontinuierlichen Saccharose-Gradienten von 0,75 – 1,5 M Saccharose in Puffer B, der von 1 ml 2,5 M Saccharose in Puffer B unterschichtet war, aufgetragen. Durch 20 min Zentrifugation bei 3000 × g sammelten sich die Zellkerne an der Grenzschicht zwischen der 2,5 M Saccharose-Unterlage und dem Gradienten. Die Kerne wurden entnommen, 10 min bei 8000 × g abzentrifugiert, in Puffer B suspendiert und auf RNA-Polymerase-Aktivität getestet.

Isolierung von RNA-Polymerase aus *Phytophthora*: Die Isolierung der RNA-Polymerase erfolgte nach der Methode von BARREAU und BEGUERET (1982). Getestet wurde die Polymerase-Aktivität (^{14}C-UTP-Einbau) im Rohhomogenat nach Aufschluß des gefrorenen Mycels in der X-Presse und nach Teilreinigung des Enzyms durch Polyethylenimin-Fällung und DEAE-Cellulose-Chromatographie.

RNA-Polymerase Test: Das Reaktionsgemisch (100 µl) enthielt 50 mM Tris-HCl (pH 7,8), 2 mM $MnCl_2$, 10 mM $MgCl_2$, je 0,3 mM ATP, GTP und CTP, 20 KBq ^{14}C-UTP, 10 µg denaturierte DNA aus Kükenerythrocyten (nicht bei Ansätzen mit Zellkernen) und 10 µl der zu testenden Lösung. Nach 60 min Inkubation bei 30 °C wurde der Ansatz auf DE81-Filterscheiben (WHATMAN) aufgetragen, mehrmals mit einer 5% Na_2HPO_4-Lösung, einmal mit dest. Wasser und zweimal mit Äthanol über einer Filternutsche gewaschen. Die Radioaktivität wurde im PACKARD Scintillationsspektrometer gemessen.

Ergebnisse

Einbau-Kinetiken von Nucleinsäure-Precursoren und Leucin

Nach Befunden von KERKENAAR und SIJPESTEIJN (1981) wird das Wachstum von *Phytophthora nicotianae* in Agar-Kultur durch Metalaxyl in Konzentrationen von 0,1 bis 0,3 µg/ml bei 1–3 Wochen Versuchsdauer zu 50% gehemmt. In unserer Submerskultur ist die Wachstumshemmung noch stärker und erreicht mit 0,3 µg Metalaxyl/ml 80%, bei 48stündiger Versuchsdauer (Tab. 1). In diesem Konzentrationsbereich er-

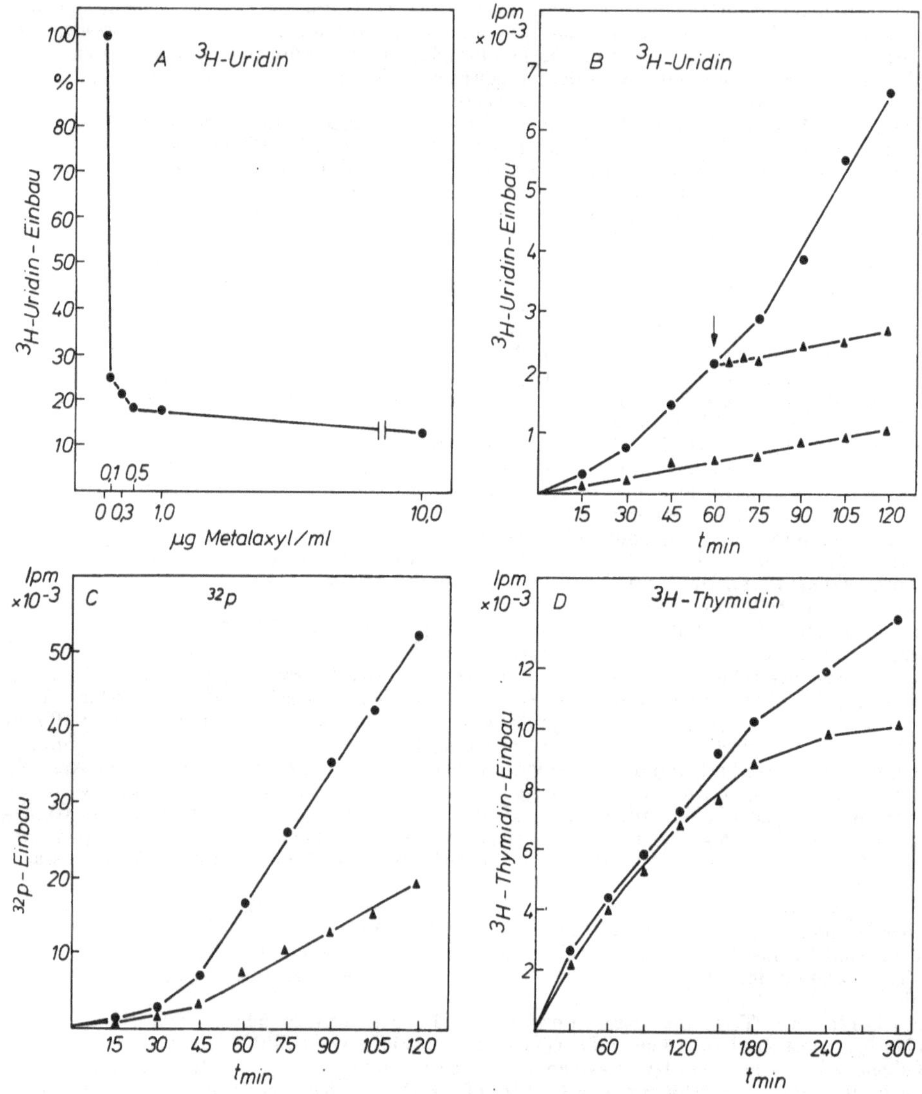

Abb. 1.
Wirkung von Metalaxyl auf den Einbau von ^{3}H-Uridin, ^{32}P und ^{3}H-Thymidin in das Mycel von *Phytophthora nicotianae*. A., Konzentrationsabhängigkeit der Metalaxylwirkung auf den ^{3}H-Uridin-Einbau. B. C.D., Zeitabhängigkeit des Einbaus von ^{32}P, ^{3}H-Uridin und ^{3}H-Thymidin. In allen Fällen wurde das Mycel vor Zugabe der Radioaktivität 15 min mit Metalaxyl vorinkubiert (in B. C. D. mit 0,5 µg/ml). In B. wurde das Fungicid zusätzlich erst 60 min nach Beginn der Inkubation mit ^{3}H-Uridin eingesetzt (Pfeil). ● Kontrolle, ▲ Metalaxyl (in B-D)

reicht auch die Hemmung des Einbaus von ^{3}H-Uridin in die RNA nahezu ihren höchsten Wert. Die maximale Hemmung wird mit 0,5 µg erreicht (Abb. 1A). Mit diesen Metalaxyl-Konzentrationen wurden alle weiteren Versuche durchgeführt. Wenn nicht anders angegeben, wurden 0,5 µg/ml eingesetzt. Der Einbau von ^{3}H-Uridin in säurefällbares Material wird während einer Versuchsdauer von 2 Std. durch Metalaxyl

Tabelle 1

Einfluß von unterschiedlichen Metalaxyl-Konzentrationen auf das Wachstum von *Phytophthora nicotianae* in Submerskultur. Die Zugabe von Metalaxyl erfolgte zum Zeitpunkt der Beimpfung (NL 963). Die Ernte erfolgte nach 48 Stunden. Trockengewicht des Inoculums 50 mg

Metalaxyl-Konzentration µg/ml NL	Trockengewicht mg/Kultur	Hemmung in %
Kontrolle	1130	0
0,1	660	42
0,3	230	80
0,5	140	88
1,0	128	89

zu über 80% gehemmt. Die Hemmwirkung setzt unmittelbar nach Zugabe des Fungicids ein (Abb. 1B), was der schnellen Aufnahme in das Mycel entspricht (FISHER u. HAYES 1982). Der Einbau von ^{32}P in säurefällbares Material wird mit 60% weniger stark gehemmt als der Uridin-Einbau (Abb. 1C), da Phosphat nicht nur in RNA, sondern auch in andere Makromoleküle (DNA, Phosphatproteide, Phospholipide) eingeht, deren Synthesen durch Metalaxyl offenbar gar nicht oder aber weniger stark beeinflußt werden als die RNA-Synthese. Der Einbau von ^{3}H-Thymidin, d. h. die DNA-Synthese, wird im Gegensatz zur RNA-Synthese in den ersten 2 bis 3 Stunden nach Metalaxyl-Zugabe nicht beeinflußt, erst dann setzt eine Hemmung ein (Abb. 1D). In ähnlicher Weise beginnt auch die Hemmung der Protein-Synthese (^{14}C-Leucin-Einbau) erst etwa 90 min nach Metalaxyl-Zugabe (60 min nach dem Leucin-Einsatz) und läuft noch über mehrere Stunden weiter, wenn auch deutlich verringert gegenüber der Kontrolle (Abb. 2A und B).

Die Aufnahme von ^{32}P, ^{3}H-Uridin, ^{3}H-Thymidin und ^{14}C-Leucin in das Mycel wird durch Metalaxyl nicht gehemmt (Ergebnisse nicht dargestellt).

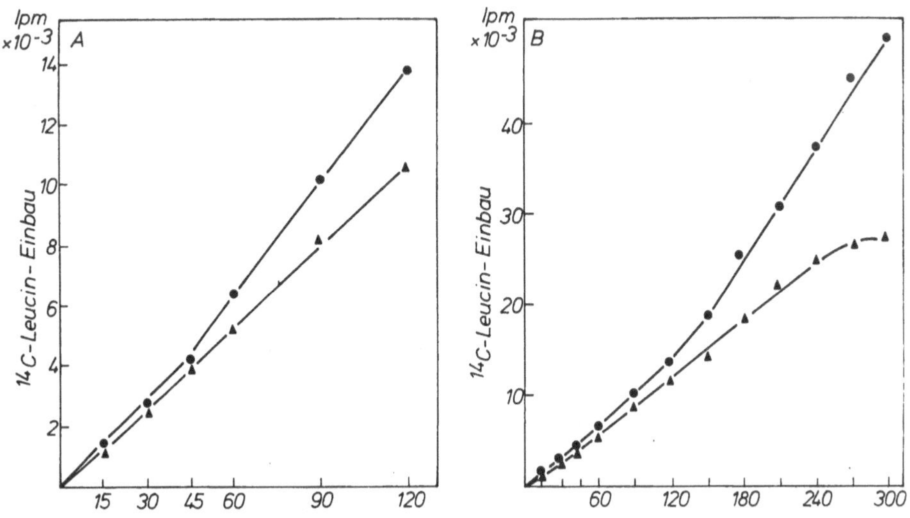

Abb. 2.
Wirkung von Metalaxyl (0,5 µg/ml) auf den Einbau von ^{14}C-Leucin in das Mycel von *Phytophthora nicotianae*. ● Kontrolle, ▲ Metalaxyl

RNA-Synthese in vivo

Da die bisher dargestellten Befunde darauf hindeuten, daß die primäre Hemmwirkung des Metalaxyl im Bereich der Transkription liegt, haben wir weitere *in vivo*- und *in vitro*-Experimente zur Charakterisierung der Hemmung der RNA-Synthese durchgeführt. Im Vordergrund der Experimente *in vivo* stand die Frage, ob die Synthese aller RNA-Arten in gleicher Weise beeinflußt wird. Um die hochmolekulare rRNA und DNA einerseits, und die niedermolekularen RNA (tRNA, 5 S rRNA) andererseits möglichst genau erfassen zu können, verwendeten wir zur elektrophoretischen Trennung zwei Polyacrylamidgel-Systeme von 2,4 und 7,5% PAA. Werden ^{32}P-markierte Nucleinsäuren auf 2,4%igen Gelen getrennt, dann zeigen die Autoradiogramme, daß nach 4 Std. Inkubationszeit die DNA-Synthese nicht beeinflußt ist, während die Synthese der rRNA wie erwartet stark gehemmt ist, und zwar bei quantitativer Auswertung zu 88% durch 0,1 µg/ml und zu 93% durch 0,5 µg Metalaxyl pro ml (Abb. 3).

Bei der ^{32}P-Markierung im Bereich der DNA handelt es sich eindeutig um eine Markierung der DNA und nicht um RNA-Moleküle, die als DNA/RNA-Komplex vorliegen könnten, denn bei der gleichen Analyse von ^{3}H-Uridin-markierten Nucleinsäuren findet sich keine Radioaktivität im Bereich der DNA (Abb. 4). Damit wird in *Phytophthora* im Gegensatz zu *Pythium splendans* (KERKENAAR 1981) Uridin spezifisch in RNA und nicht auch in DNA eingebaut. Der Einbau von ^{3}H-Uridin in die rRNA wird mit über 90% (Abb. 4) in der gleichen Weise gehemmt wie der Einbau von ^{32}P (Abb. 3).

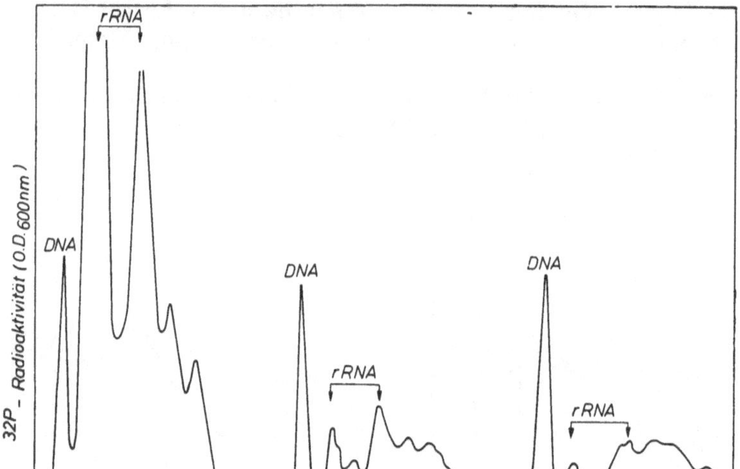

Abb. 3.
Wirkung von Metalaxyl auf den Einbau von ^{32}P (3 Stunden) in ribosomale RNA und DNA von *Phytophthora nicotianae*. Die Nucleinsäuren (je 30 µg) wurden in 2,4% PAA-Gelen getrennt, die Lage der Nucleinsäure-Fraktionen durch Scannen der optischen Dichte bei 260 nm ermittelt. Die Autoradiogramme der getrockneten Gele wurden densitometrisch ausgewertet

Eine Analyse der gleichen radioaktiv markierten RNA auf 7,5% PAA-Gelen zur Auftrennung der niedermolekularen RNA ergab, daß die Synthese der 5 S rRNA durch Metalaxyl in gleicher Weise gehemmt wird wie die Synthese der hochmolekula-

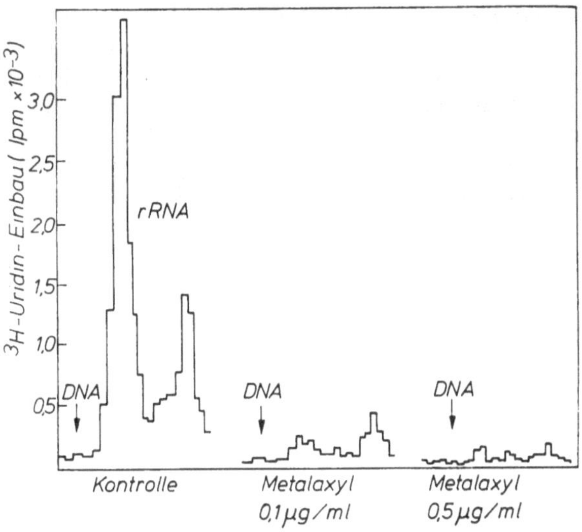

Abb. 4.
Wirkung von Metalaxyl auf den Einbau von ^{3}H-Uridin in die ribosomale RNA von *Phytophthora nicotianae*. Analyse der Nucleinsäuren wie in Abb. 3. Die Radioaktivität wurde nach Schneiden der Gele in 2 mm Fraktionen ermittelt

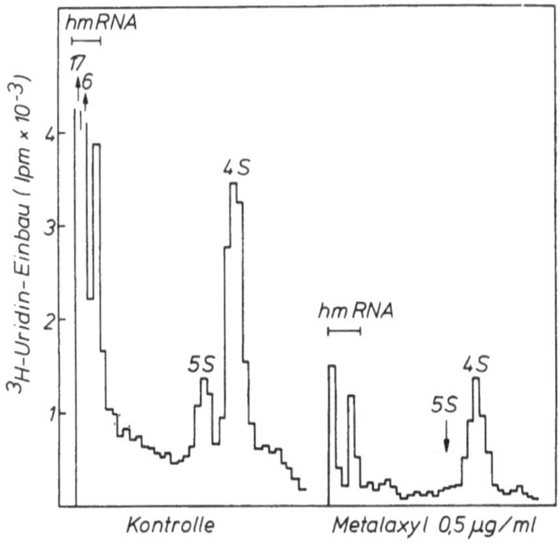

Abb. 5.
Wirkung von Metalaxyl auf den Einbau von ^{3}H-Uridin in die niedermolekulare RNA (5 S rRNA, 4 S tRNA) von *Phytophthora nicotianae*. Die Nucleinsäuren (je 40 µg) wurden auf 7,5 % PAA-Gelen getrennt, die Lage der Nucleinsäuren durch Scannen der optischen Dichte bei 260 nm und durch zusätzliche Elektrophorese gereinigter tRNA ermittelt. In der Startregion verbleibt die hochmolekulare RNA (hmRNA). Die Radioaktivität wurde nach Schneiden der Gele in 2 mm Fraktionen ermittelt

ren RNA. Andererseits ist der ^{3}H-Uridin-Einbau in die 4 S RNA (tRNA) mit etwa 55% wesentlich weniger stark gehemmt (Abb. 5). Die gleiche Analyse (Abb. 5) zeigt außerdem, daß der Uridin-Einbau in die in der Startregion der Gels verbleibende hochmolekulare RNA entsprechend den Befunden in Abbildung 4 zu etwa 90% gehemmt ist. Eine Analyse der ^{32}P-markierten RNA ergab die gleichen Befunde (nicht dargestellt). In beiden Fällen verändert sich beim Auftragen von Gesamt-Nucleinsäuren auf das Gel das Verhältnis der Radioaktivität in rRNA zu tRNA von 6–8:1 in der Kontrolle auf etwa 2:1 in der mit Metalaxyl behandelten Pilzkultur.

Tabelle 2

Wirkung von Metalaxyl auf den Einbau von ^{32}P in die Poly(A)$^+$-RNA von *Phytophthora nicotianae*. Die Werte bedeuten den Anteil Radioaktivität in der Poly(A)$^+$-RNA an der Gesamt-Radioaktivität der hochmolekularen RNA

	Experiment 1	Experiment 2
Kontrolle	2,2%	2,5%
Metalaxyl (0,5 µg/ml)	9,3%	8,4%

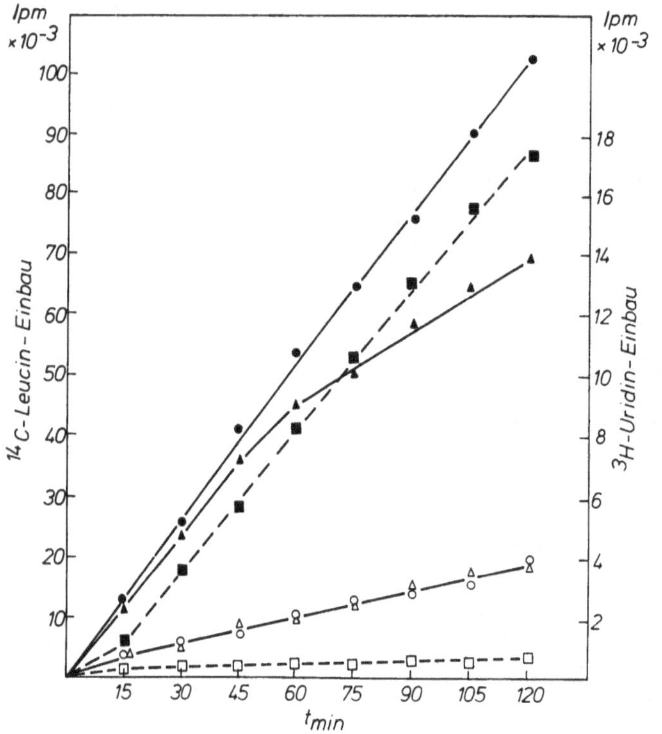

Abb. 6.

Wirkung von Metalaxyl (0,5 µg/ml) und Actinomycin (50 µg/ml) auf den Einbau von ^{14}C-Leucin in das Protein und von ^{3}H-Uridin in die RNA von *Phytophthora nicotianae*. Das Mycel wurde vor Zugabe der Radioaktivität 30 min mit den Hemmstoffen vorinkubiert. ^{14}C-Leucin: ● Kontrolle, ▲ Metalaxyl, ○ Actinomycin, △ Metalaxyl + Actinomycin. ^{3}H-Uridin: ■– – –■ Kontrolle, □ – – □ Actinomycin

Der Einbau von ^{32}P oder ^{3}H-Uridin wird ähnlich wie in die tRNA auch in die Poly(A)$^{+}$-RNA (mRNA) durch Metalaxyl weniger gehemmt als der Einbau in die rRNA. Bedingt durch die Aufarbeitungsmethode ist eine quantitative Bestimmung von Poly(A)$^{+}$-RNA schwierig. Deshalb wählten wir als Maß für die radioaktive Markierung von Poly(A)$^{+}$-RNA in Kontrolle und Metalaxyl-behandelten Kulturen den Anteil der Radioaktivität in der Poly (A)$^{+}$-RNA (in%) an der Gesamtmarkierung der hochmolekularen RNA. Die hochmolekulare RNA (rRNA und mRNA) gewannen wir durch Fällung mit 3 M LiCl aus Gesamtnucleinsäure-Präparationen. Tabelle 2 enthält die Ergebnisse von 2 Experimenten. Die starke relative Zunahme von markierter Poly(A)$^{+}$-RNA in der markierten hochmolekularen RNA unter Metalaxyl-Einfluß zeigt, daß die Synthese der Poly(A)$^{+}$-RNA durch das Fungicid wesentlich weniger gehemmt wird als die Synthese der ribosomalen RNA.

Die bisher dargestellten Ergebnisse zeigen einerseits, daß Metalaxyl die rRNA-Synthese stärker hemmt als die Synthese von tRNA und mRNA, und sie deuten andererseits darauf hin, daß der primäre Hemmeffekt in der RNA-Synthese zu suchen ist bei sekundärer Beeinflussung der Proteinsynthese. Das in Abbildung 6 dargestellte Experiment unterstützt diese Annahme. Actinomycin als Inhibitor der Transkription hemmt die RNA-Synthese vollständig, aber auch die Protein-Synthese wird sehr schnell weitgehend blockiert. Metalaxyl hat dabei keinen zusätzlichen Einfluß. Somit ist der Hauptteil der mRNA in *Phytophthora* kurzlebig, bei fehlender Neusynthese in Gegenwart von Actinomycin ist deshalb die Protein-Synthese gehemmt. Da Metalaxyl allein die Protein-Synthese erst wesentlich später und schwächer hemmt als Actinomycin, muß in Gegenwart von Metalaxyl noch ausreichend mRNA synthetisiert werden, um diese Protein-Synthese zu ermöglichen. Dieser Befund bestätigt die in Tabelle 2 dargestellten analytischen Werte.

RNA-Synthese in vitro

Die unterschiedliche Hemmung der Synthese der einzelnen RNA-Arten ließ uns vermuten, daß Metalaxyl die Aktivität der RNA-Polymerasen, bevorzugt der Polymerase I, beeinflußt. Eine Wirkung auf der Ebene des Chromatins oder der DNA-Matrize war aber ebensowenig auszuschließen. Die als vorläufig zu wertenden Ergebnisse der in Tabelle 3 dargestellten Experimente zeigen, daß die Aktivität einer teilgereinigten

Tabelle 3

Wirkung von Metalaxyl auf die RNA-Synthese (^{14}C-UTP-Einbau) durch ein Zellhomogenat, eine teilgereinigte RNA-Polymerase und zwei Zellkernfraktionen von *Phytophthora nicotianae*

	Zellhomogenat		Teilgereinigte RNA-Polymerase		Zellkern-Rohfraktion		Zellkerne gereinigt	
	I. p. m.	%	I. p. m.	%	I. p. m.	%	I. p. m.	%
Kontrolle	1627	100	284	100	2730	100	1808	100
Metalaxyl (0,1 μg/ml)	822	50	298	104	1934	71	1246	69

RNA-Polymerase-Präparation durch Metalaxyl nicht beeinflußt wird, obwohl das Enzym durch Actinomycin und α-Amanitin in normaler Weise gehemmt wird. Wir verzichten deshalb auf eine weitere Reinigung und Fraktionierung der Polymerasen. Die geringe Aktivität des Enzyms erklärt sich aus der starken Verdünnung im Säuleneluat nach DEAE-Cellulose-Chromatographie. Im Unterschied zum gelösten Enzym wird die RNA-Synthese im Zellhomogenat und in Zellkernfraktionen durch Metalaxyl signifikant gehemmt.

Wirkung von Metalaxyl auf die RNA-Synthese in pflanzlichen Systemen

Untersuchungen anderer Autoren haben ergeben, daß die Wachstumshemmung durch Metalaxyl und andere Acylalanin-Verbindungen auf die Peronosporaceen beschränkt zu sein scheint. An einigen in unserem Labor bearbeiteten Systemen überprüften wir, ob die RNA-Synthese höherer Pflanzen durch Metalaxyl beeinflußt werden kann. In Zellkulturen der Solanacee *Lycopersicon peruvianum* wird der ^{3}H-Uridin-Einbau in die RNA bis zu einer getesteten Konzentration von 10 µg Metalaxyl pro ml nicht beeinflußt (Methoden nach Scharf u. Nover 1983). Ebenso wird der ^{14}C-UTP-Einbau in die RNA einer Zellkernfraktion, die nach Ultraschall-Aufschluß der Tomaten-Zellkulturen gewonnen wurde, durch die gleichen Metalaxylkonzentrationen nicht gehemmt bei einer Hemmung von 42% durch 1 µg α-Amanitin pro ml. Auch die Aktivität einer hochgereinigten RNA-Polymerase aus Spinat-Chloroplasten (Methode nach Lerbs *et al.* 1983) wird durch 0,1 bis 1 µg Metalaxyl pro ml nicht gehemmt.

Diskussion

Die Geschwindigkeit der Wirkung sehr niedriger Metalaxyl-Konzentrationen auf das Mycelwachstum, zumal bei Berücksichtigung der geringen Aufnahme des Fungicids durch das Mycel (Kerkenaar 1981, Fisher u. Hayes 1982), spricht für eine sehr spezifische Wirkung des Hemmstoffs. Die Geschwindigkeit der Hemmwirkung auf die RNA-Synthese im Vergleich zu DNA- und Protein-Synthese, und die Hemmung der RNA-Synthese in isolierten Zellkernen weisen darauf hin, daß der oder zumindest ein wesentlicher Angriffspunkt des Metalaxyls im Bereich des RNA-Stoffwechsels liegt. Die stark verzögert einsetzende Hemmung der Protein-Synthese ist sicher sekundärer Natur. Auch in Gegenwart von Metalaxyl bleibt die Protein-Synthese über längere Zeit intakt, da die Ribosomen so stabil sind, daß es trotz nahezu totaler Hemmung der rRNA-Synthese zunächst zu keinem Mangel an Ribosomen kommt. Andererseits ist die Hemmung der Synthesen von tRNA und mRNA wesentlich geringer als die Hemmung der rRNA-Synthese, so daß trotz geringer Lebensdauer der mRNA die Protein-Synthese über einen gewissen Zeitraum erhalten bleibt. Der molekulare Angriffspunkt des Metalaxyl auf die Transkription ist noch ungeklärt. Die Tatsache, daß die Aktivität der freien RNA-Polymerase durch Metalaxyl nicht beeinflußt wird, muß nicht bedeuten, daß sie auch in der Zelle als Angriffspunkt auszuschließen ist. Es wäre durchaus denkbar, daß bei der Polymerase-Präparation Bindungsfaktoren für das Fungicid verloren gehen. Hierzu sind weitere Untersuchungen erforderlich. Es bleibt ebenso offen, worauf die unterschiedlich starke Hemmung der Synthese der drei RNA-Arten beruht. Für eine denkbare unterschiedliche Affinität der drei Polymerasen für Metalaxyl gibt es noch keine Anhaltspunkte. Alternative Möglichkeiten für Mechanismen der Transkriptionshemmung wären Eingriffe auf Chromatinebene über Regulationsproteine, oder aber eine Bindung an die DNA, eventuell unterschiedlich stark an unterschiedliche DNA-Bereiche, die für die verschiedenen RNA-Arten codieren.

Die Hemmung der DNA-Synthese durch Metalaxyl setzt in *Phytophthora* im Gegensatz zu Befunden an *Pythium splendans* (Kerkenaar 1981) wesentlich später ein als die der RNA-Synthese und auch noch verzögert gegenüber der Protein-Synthese-Hemmung. Das deutet darauf hin, daß die Hemmung der DNA-Synthese und auch der Mitose (Fisher u. Hayes 1982) Sekundärwirkungen sind, die auf einer Hemmung der RNA-Synthese beruhen. Unsere Ergebnisse über die Wirkung von Metalaxyl auf die RNA-Synthese *in vivo* und *in vitro* stimmen weitgehend mit den kürzlich von Davidse (1983) mitgeteilten Befunden an *Phytophthora megasperma* überein.

Anmerkungen bei der Korrektur: Nach Befunden von DAVIDSE, GERRITSMA und HOFMAN (Phytiatrie-Phytopharmacie **30**, 235–244, 1981) wird auch in *Phytophtora megasperma* der Uridin-Einbau in die RNA sofort nach Metalaxylzugabe gehemmt bei nur geringer Hemmung der DNA- und Proteinsynthese. Die Uridin-Aufnahme ist nicht beeinflußt.

Wir danken Herrn Prof. Dr. H. LYR, Kleinmachnow, für die Überlassung von Metalaxyl und des Stammes von *Phytophthora nicotianae*. Den Herren Prof. Dr. B. PARTHIER und Prof. Dr. D. GRÖGER danken wir für vielfältige Diskussionen und Herrn Dr. L. NOVER für die Zellkulturen von Tomate und für die Unterstützung bei den Versuchen mit Zellkulturen. Frau S. SCHWARZ und Frau I. REEH sind wir sehr dankbar für die zuverlässige technische Mitarbeit.

Literatur

BARREAU, C. and BEGUERET, J., 1982. DNA-dependent RNA polymerase III from the fungus *Podospora comata*. Purification, subunit structure and comparison with the homologous enzyme of a related species. Eur. J. Biochem., **129**, 423–428.

BENSON, D. M., 1979. Efficacy and *in vitro* activity of two systemic acylalanines and ethazole for control of *Phytophthora cinnamomi* root rot of azalea. Phytopathology, **69**, 174–182.

DAVIDSE, L. C., 1983. Mechanism of action and resistance to metalaxyl. Vortrag auf der Tagung „Systemic Fungicides and Antifungal Compounds" Reinhardsbrunn, DDR, 20–25. Mai 1983 (Abstract).

FISHER, D. J. and HAYES, A. C., 1982. Mode of action of the systemic fungicides furalaxyl, metalaxyl and ofurace. Pestic. Sci., **13**, 330–339.

FRIES, N., 1938. Über die Bedeutung von Wuchsstoffen für das Wachstum verschiedener Pilze. Symbolae Bot. Upsaliensis III, 2.

KERKENAAR, A., 1981. On the antifungal mode of action of metalaxyl, an inhibitor of nucleic acid synthesis in *Pythium splendans*. Pestic. Biochem. Physiol. Pflanzen, **16**, 1–13.

KERKENAAR, A. and SIJPESTEIJN, H. K., 1981. Antifungal activity of metalaxyl and furalaxyl. Pestic. Biochem. Physiol., **15**, 71–78.

LERBS, S., BRIAT, J.-F. and MACHE, R., 1983. Chloroplast RNA polymerase from spinach: purification and DNA-binding proteins. Plant Mol. Biol. (in press).

MUNSCHE, D. und WOLLGIEHN, R., 1973. Die Synthese von ribosomaler RNS in Chloroplasten von *Nicotiana rustica*. Biochim. biophysica Acta, **294**, 106–117.

SCHARF, K.-D. and NOVER, L., 1983. Heat-shock-induced alterations of ribosomal protein phosphorylation in plant cell cultures. Cell, **30**, 427–437.

SCHIEMANN, J., WOLLGIEHN, R. and PARTHIER, B., 1978. DNA-dependent RNA polymerase in *Euglena gracilis* broken chloroplasts. Biochem. Physiol. Pflanzen, **172**, 507–519.

SCHWINN, F. J., STAUB, T. und URECH, P. A., 1977. Die Bekämpfung falscher Mehltau-Krankheiten mit einem neuen Wirkstoff aus der Gruppe der Acylalanine. Mitt. Biol. Bunds. Anst. Ld.- u. Forstw. Berlin, **178**, 145–146.

STAUB, T. H., DAHMEN, H. und SCHWINN, F. J., 1978. Biological characterization of uptake and translocation of fungicidal acylalanines in grape and tomato plants. Z. Pflanzenkrankh. Pflanzenschutz, **85**, 162–168.

STAUB, T. H., DAHMEN, H. and SCHWINN, F. J., 1980. Effects of ridomil on the development of *Plasmopara viticola* and *Phytophthora infestans* on their host plants. Z. Pflanzenkrankh. Pflanzenschutz, **87**, 83–91.

URECH, P. A., SCHWINN, J. and STAUB, T., 1977. CGA 48988, a novel fungicide for the control of late blight, downy mildews and related soil borne diseases. Proc. Brit. Crop Prot. Conf. Brighton, 623–632.

WOLLGIEHN, R., LERBS, S. und MUNSCHE, D., 1978. Eigenschaften einer Chloroplasten-RNA vom Molekulargewicht 0.5×10^6 aus *Nicotiana rustica*. Biochem. Physiol. Pflanzen, **173**, 60–69.

WOLLGIEHN, R., MUNSCHE, D. und MIKULOVICH, T. P., 1974. ^{32}P-Inkorporation in die Chloroplasten-RNS von Tabakblättern. Ausschluß der Beteiligung von Bakterien-Verunreinigungen. Biochem. Physiol. Pflanzen, **165**, 37–48.

Anschrift: Dr. habil. R. WOLLGIEHN, Akademie der Wissenschaften der DDR, Institut für Biochemie der Pflanzen, DDR-4020 Halle/Saale, Weinberg 3

Zeitschrift für allgemeine Mikrobiologie 24 (1984) 4, 280

Buchbesprechung

R. PIECHOKI, Genmanipulation — Frevel oder Fortschritt? 228 S., 8 Tab. 42 Abb. Leipzig-Jena-Berlin 1983. Urania-Verlag. M 12,80.

Wohl keine umwälzende Entwicklung eines Wissenschaftsgebietes ist von soviel publizistischer Aktivität begleitet worden wie die Erarbeitung und Anwendung der modernen Methoden der Gentechnik und die genetische Manipulation höherer Organismen durch gezielte Eingriffe in die Fortpflanzungsmechanismen. Das vorliegende Buch — „Genmanipulation — Frevel oder Fortschritt?" — reiht sich damit in eine Vielzahl von publizistischen Beiträgen zu dieser Thematik ein. Sein Wert liegt darin, daß die modernen Ergebnisse biologischer Forschung einer breiten Öffentlichkeit zur Meinungsbildung bekannt gemacht werden.

Nach einleitenden Kapiteln, die den interessierten Laien mit dem notwendigen Grundlagenwissen zum Verständnis der Problematik ausstatten sollen, werden wesentliche Methoden der Gentechnik und deren erste spektakuläre Ergebnisse dargestellt. Dabei war der Autor — selbst Biologe und kein Wissenschaftsjournalist — bemüht, eine komplizierte Problematik in verständlicher Form abzuhandeln. Daß dies nicht immer in klarer und übersichtlicher Form gelungen ist, unterstreicht die Schwierigkeit, komplizierte wissenschaftliche Methoden und Ergebnisse aus dem Fachjargon heraus in eine jedem verständliche Sprache zu übersetzen. Hilfreicher wäre auch ein geringerer Informationsgehalt der einzelnen Schemata, ihre Ausführung im Mehrfarbendruck und schließlich eine bessere Kopplung von Schema und zugehörigen Textstellen gewesen.

Weitere Kapitel fassen die Entwicklung auf dem Gebiet der genetischen Manipulation von Pflanzen und Tieren zusammen, und schließlich werden humangenetische Fragen und die heutigen Möglichkeiten zum gezielten Eingriff in das Erbgut und die Fortpflanzung des Menschen dargestellt. Damit beschränkt sich das Buch nicht auf die Genmanipulation an sich, sondern die ganze Palette genetischer Manipulationen, sowohl auf molekularem wie auch auf dem Fortpflanzungsniveau, wird in erfreulicher Gesamtheit behandelt.

Die die Entwicklung der Gentechnik begleitende publizistische Welle hat sich besonders mit ethischen Fragen, Fragen der Verantwortung des Wissenschaftlers u. ä. auseinandergesetzt. Auch das vorliegende Buch geht ausführlich darauf ein. Leider stellt aber bereits der Untertitel des Buches — Frevel **oder** Fortschritt — dem Leser eine Alternative, die so nicht stehen kann. Am Ende der Lektüre des Buches muß beim Leser ein negativer Gesamteindruck entstehen. Zuviel Raum wird Horrorvisionen eingeräumt, die längst zum größten Teil durch fundierte wissenschaftliche Ergebnisse widerlegt sind. Unter den möglichen Gefahren und Konsequenzen der Gentechnik muß die Vernichtung von Arbeitsplätzen in Entwicklungsländern als etwas weit hergeholt anmuten (S. 140). Auch werden wohl die Verhältnisse sehr tendenziös dargestellt, wenn man den Gentechnikern als Hauptmotiv für ihre Forschung „... Anerkennung, Karriere und finanzielle Gewinne" unterstellt, humanistische Motive aber als untergeordnet eingestuft werden. Gerade der zeitweilige freiwillige Verzicht auf bestimmte Experimente und die Erarbeitung von Sicherheitsvorschriften durch die beteiligten Wissenschaftler sind wohl bisher einmalige Ergebnisse, die das Verantwortungsbewußtsein der Wissenschaftler deutlich belegen.

In einer Neuauflage des Buches sollten sinnentstellende Druckfehler (S. 28 u. 97: Milliliter statt Millimeter) und die zwar im Text angekündigte aber fehlende Tabelle (S. 172) zur Häufigkeit genetisch bedingter Defekte bei Neugeborenen korrigiert bzw. eingefügt werden.

D. BEHNKE (Jena)

ZEITSCHRIFT FÜR ALLGEMEINE MIKROBIOLOGIE

VOLUME 24 1984 NUMBER 4

CONTENTS

Zeitschrift für allgemeine Mikrobiologie is indexed in Current Contents/Life Sciences and reviewed by Chemical Abstracts.

FSC
www.fsc.org
MIX
Papier | Fördert
gute Waldnutzung
FSC® C083411

Zeitfracht Medien GmbH
Ferdinand-Jühlke-Straße 7
99095 Erfurt, Deutschland
produktsicherheit@kolibri360.de